AF555922

CHINESE ABACUS
Suopan Techniques

CHINESE ABACUS

Suopan Techniques

A.S. BHATIA

REGAL PUBLICATIONS
New Delhi-110027

CHINESE ABACUS
Suopan Techniques

ISBN 978-81-8484-355-2

Typeset by
S.S. COMPOSERS
3190, Mohindra Park, Shakur Basti, Delhi-110034.

Printed in India at
MAYUR ENTERPRISES
WZ Plot No. 3, Gujjar Market, Tihar Village, New Delhi-110018.

Published by
REGAL PUBLICATIONS
F-159, Rajouri Garden, New Delhi-110027.
Phone: +91-11-45546396
E-mail: regalbookspub@yahoo.com

Dedicated to my grandchildren
Ronak** and **Sahib
for their love and affection.

Contents

Preface

The abacus is basically mechanically aided tool used for counting and forming basic arithmetic; though it is not a calculator it is a kind of calculator, similar to the four function calculators that are used in many elementary school classrooms. Whatever is available for written arithmetic can be solved by abacus.

This book describes some techniques for manipulating it for arithmetic. The Chinese techniques are described in this book.

But whether a Chinese or a Japanese version (see Japanese book) with the bead or rod abacus many centuries long calculations could be done much more quickly than on paper. This typical Asian speciality came to end with the spreading of electronic calculators. But a final rear up against the new technology was when the abacus won over the electric calculator in an exciting contest at 1946 in Tokyo.

It is easy for the beginner to learn—The notation system that each rod stands for a definite digit and each bead counter, a specific number, leaves no doubt whatsoever even in the mind of the very beginner. Especially when one is to carry out the operations for multiplication and division, one can master the methods in a blink of an eye provided that one has an understanding on the methods employed in ordinary arithmetic, and do a much faster and better job than to write down all the steps on paper.

It is light weight easy to operate on and has the following features: It has a vivid structure, an order of digits, free parallel arrangement of bead-codes and easy variety of numbers. The number is expressed with beads moving. Addition and subtraction are connected to each other. The four fundamental operations are in close connection and interchange, with clear arithmetic principles.

Abacus and mental arithmetic, without moving heads, has a much higher operation speed than abacus, both fast and convenient and with a higher seed and accuracy than written arithmetic, mental arithmetic by image of abacus can reduce calculation burdens in student's assignments to a great extent.

Those who often calculate in mental arithmetic by image of abacus have a cute sense of numerals and supper work efficiency.

Very efficient Suopan (Chinese) and Soroban (Japanese) techniques have been developed to do multiplication, division, square root and cube root operations at high speeds. The Chapter on square roots is described in details only in Japanese book. It can also be studied for Chinese abacus.

More briefly explained, the Abacus is an instrument to augment mental arithmetic. The mechanics of its use is simple, arid its accuracy is assured by repetition and practice alone.

A.S. BHATIA

Introduction

Many pages have been written about the abacus. The abacus appears in different guises in many countries.

During the 11th century the Chinese abacus, or suanpan was invented. The Suanpan is generally regarded as the earliest abacus with beads on rods. A suanpan has 2 beads above a middle divider called a **beam** and 5 beads below.

Here is a picture of a traditional Chinese Abacus. As you can see this instrument has 2 beads above the reckoning bar and 5 beads below it.

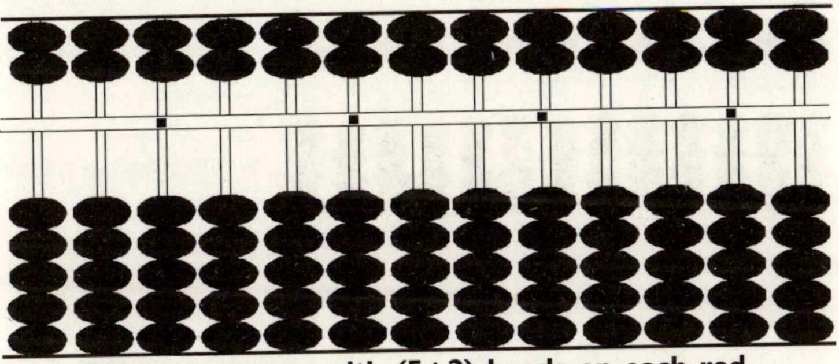

Chinese suanpan with (5+2) beads on each rod.

And below here is a picture of a traditional Japanese Soroban. It is a streamlined instrument with 1 bead above the reckoning bar and 5 below it.

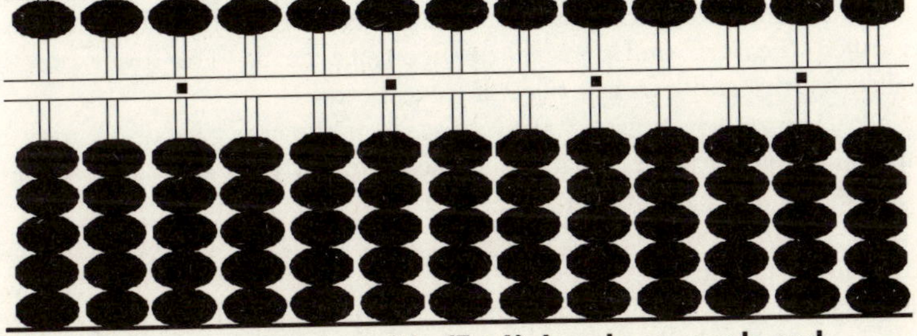

Japanese soroban with (5+1) beads on each rod.

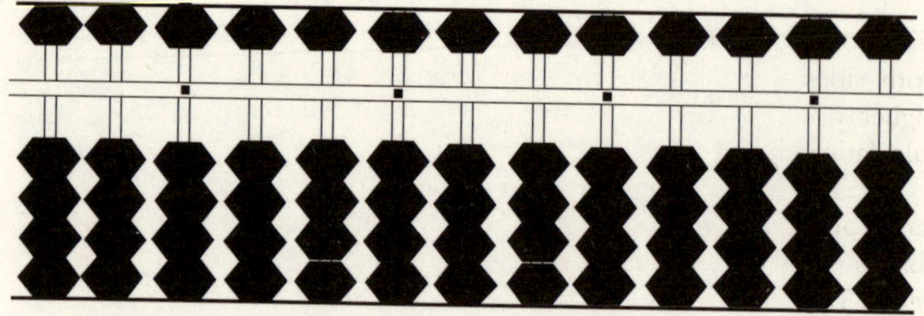

Japanese soroban with (4+1) beads on each rod.

Use of the suanpan spread to Korea, and then to Japan during the latter part of the 15th century. The Japanese termed the abacus a soroban. Originally the soroban looked very much like its Chinese but is having two beads above the reckoning bar and five beads below. Around 1850, it was modified to have only one bead above the reckoning bar while maintaining the five beads below. It was further changed by removing one lower bead in 1930. This one bead above four beads below (1/4) arrangement remains as the present day Japanese soroban construction. (see Figure above)

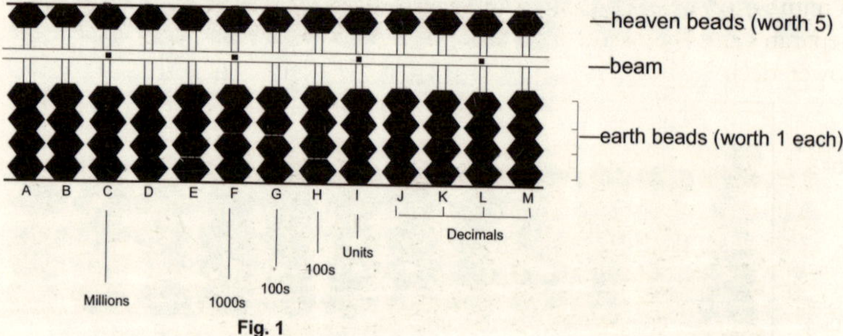

Fig. 1

A soroban is made up of a frame with vertical rods on which beads move up and down. Dividing the upper and lower portion of the soroban is a horizontal bar called a **beam** or **reckoning bar**. (see Fig. 1)

A crosswise bar divides the abacus into two parts. The upper part is called "heaven" and consists of five-value beads. The lower part is called 'earth" and consists of one-value beads.

The zero-position of all abacuses is, if the upper beads are moved up and the lower beads are moved down. Only if one or more beads are moved toward to the beam, than they represent a number.

In principle the rods represent (starting from the right side and continuing to the left side) the "ones", the 'tens", the "hundreds" etc. The user has to decide which rod the "ones" should be. If there are numbers with values less than one, it is convenient to put the column with "ones" just in the middle of the abacus. So numbers can grow to both sides. But if there are only numbers with values from one and higher you can start just at the right side of the frame developing the calculation in direction of the left side.

The Japanese abacus has two decks of beads on each wire one rank contains four beads, and the other a single bead. Like the Japanese abacus (See Japanese abacus book), the Chinese abacus has two decks of beads The lower deck has five "earth beads" on each wire, while the upper deck has two "heaven beads" on each wire.

However, the Chinese have their own version of the abacus, a 2:5 bead suanpan. There are those who prefer to use a Chinese instrument because it has a larger frame and larger beads allowing for larger fingers. It really doesn't matter which instrument you use. The procedures are virtually the same for both and these methods are well suited to either instrument.

The abacus is typically constructed of various types of hardwoods and comes in varying sizes. Most often constructed of a wood frame with beads sliding on wire or wooden pegs.

The frame of the abacus has a series of vertical rods on which a number of wooden beads are allowed to slide freely. A horizontal beam separates the frame into two sections, known as the upper deck and the lower deck (Figure 2).

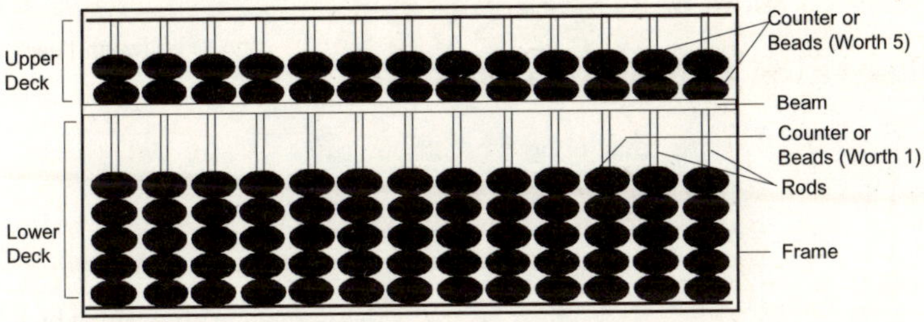

Fig. 2: Abacus Parts: The various parts of the abacus are identified here: the frame, the beam, the beads and rods and the upper and lower decks.

An Chinese abacus is divided into two parts by a long divider that stretches horizontally across the frame. Above this divider is "Heaven" where two beads reside of each vertical peg. Each bead in "Heaven" represents a count of five (5) units. Below the divider resides "Earth" where 5 beads reside on each vertical peg. Each bead resents one (1) unit Division also occurs vertically on the abacus. Each vertical peg represents a numerical sequence of 'tenths". For example, reading left to right, the first vertical peg represents one unit, the second ten units, the third one hundred units, the fourth one thousand units and so on.

Unit points on the crosspiece are usually placed one every fifth upright.

Unit points are used to indicate "ones", "thousands", "millions". "billions" and so on.

Basics

The abacus is prepared for use by placing it flat on a table or one's lap and pushing all the beads on both the upper and lower decks away from the beam. The beads are manipulated with either the index finger or the thumb of one hand.

The Beam

On a modern-day soroban one bead sits above the beam and four beads sit below. The beads above the beam are often called heaven beads and each has a value of 5. The beads below are often called earth beads and each has a value of 1. Beads only have meaning when they are pushed as far as they can go to the center of the abacus, the beam that divides the ones from the fives. Along the length of the beam, you'll notice that *every third rod is marked with a dot.* These specially marked rods are called unit rods because any one of them can be designated to carry the unit number. While the soroban operator makes the final decision as to which rod will carry the unit number, it is common practice to choose a unit rod just to the right of center on the soroban.

How to Use an Abacus Step-by-Step

1. Lay the abacus horizontally in front of you with the side containing only one bead at the top, furthest away from you, and the side with four beads at the bottom, closest to you. The top beads stand for multiples of five, and the bottom beads stand for multiples of one.
2. Enter a number into the abacus by starting on the right side using each row of beads to stand for one number. For example, to enter 1,254 the furthest row on the right would be the number 4, the row to the left of it would be 5, the row to the left of that would be 2, and the row to the left of that would be 1.
3. Enter the numbers by sliding the appropriate number of beads from the bottom up the bar in the middle. For the number 4, you would slide all 4 beads up. For 2 you would slide 2 beads up, and so on.
4. Enter the number five by finding the appropriate row and slide the one single bead at the top down. The single beads that are above the bar in the middle count up starting with

5 on the far right. The next bead would be 10, then 15, and so on.

5. Number 3 to 1,254 you would start with the row where you have entered the number 4. Since you do not have single beads left to add you add one which turns 4 into 5. To show this, you slide the 4 beads back to their original position and slide the single 5 bead down. Then slide 2 beads back up. You have just added 4+3 to make 7.
6. Move down the rows adding numbers. If a number becomes greater than 10, you move on to the next row. For example if you add 7 to 1254 you start with 4. Slide the 4 beads back down and slide the 5 bead down. Then push 4 beads back up. This gives you 9 and you have used all your beads, but you have only added 4+5. You still need 2 more to make 4+7. These 2 get added to the next row to the left. If this row adds up to be more than 9, you add what is left over on to the next row.
7. Continue moving across the rows until you have finished adding. To subtract, you follow the same procedure as addition, only moving from left to right instead of right to left.

The dots also serve as markers by which larger numbers can be quickly and efficiently recognized. For example in Fig. below (3a) rod I is the designated unit rod. For example, using K as unit rod when given a number such as 12, 345, 6789 an operator can quickly identify rod C as the 100 millions rod and go ahead and set the first number on that rod. This ensures that all subsequent numbers will be set on their correct rods and that the unit number 9 will fall neatly on unit rod K as shown in Figure (3b).

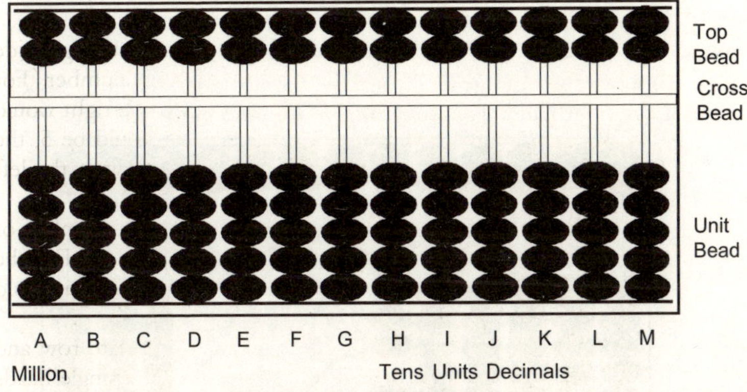

Fig. 3a: Chinese Abacus

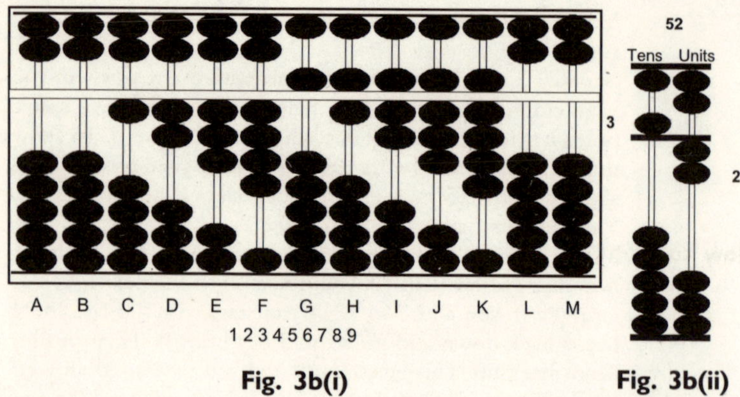

Fig. 3b(i) **Fig. 3b(ii)**

If a dot is shown at I. It is ready to register numbers from 0 to 9, the second rod to its left H is to be used for registering all digits of the order of tens, the third rod to it is left G is to be used for registering all digits of the order of hundreds, etc. The rods to the right of the dot are to be used to register digits of the order of tenths, hundredths, thousandths, etc.

To register any given number on the abacus, remember: Each bead counter of the lower deck has the value of 1, while each bead counter of the upper deck has the value of 5. The number registered on the abacus in the above (Fig. 3b) should therefore read 12, 345, 6789.

The Way to Move the Bead

To use the abacus it is essential to aim at speed and accuracy, Proper finger technique is paramount in achieving proficiency on the abacus. Remarkable skill can be attained if the user is able to achieve coordination of his fingers, and his hand and brain. Accuracy and speed largely depend on skilful manipulation of the bead counters by the fingers.

To move the bead generally use the thumb, index finger or middle finger of the right hand. The three fingers are used in the following way:

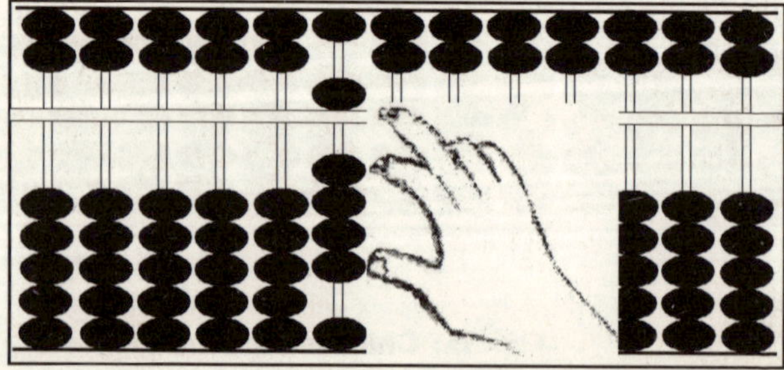

(1) The thumb controls the movement of the lower bead counters upwards to the beam.
(2) The index finger controls the movement of the lower bead counters downwards away from the beam.
(3) The middle finger controls the movement of the upper bead counters downwards to the beam, or away from the beam.

How to Manipulate your Fingers

In pushing the counters, there are three ways.

(1) Use only the forefinger.
(2) Using the thumb and forefinger.
(3) Using the thumb, forefinger and middle finger.

The third one is the method with the thumb.

Of these three ways, the second one using the thumb and forefinger is recommended it being the best and easiest.

"Take off" a counter in the upper part. -

Note: "Put in" means pushing the counters towards the crosspiece.

"Take off" means pushing the counters away from the crosspiece

Thus, with a Chinese abacus, the thumb and the index finger together with the middle finger are used to manipulate the beads. Beads in lower deck are moved up with the thumb and down with the index finger. In certain calculations, the middle finger is used to move beads in the upper deck.

Bead Values

Each bead in the upper deck has a value of 5; each bead in the lower deck has a value oft. Beads are considered counted, when moved towards the beam that separates the two decks. Move all the beads to the bottom. To read the abacus lay the abacus on a flat surface. Position the abacus so the majority of the beads—the "1" beads—are closest to you and the "5" beads are away from you. Press all beads to the upper or lower frame boards, respectively—away from the middle space.

How to Manipulate your Counters

The manipulation consists of the following six motions.

(1) Putting in counters.
(2) Taking off 'counters.
(3) Putting in five counter and taking off one-counter.
(4) Putting in one-counter and takin off five counter.

(5) Taking off counter of ones' place and putting in a counter of tens place.
(6) Taking off a counters of tens' place and putting in counters of ones' place:

Note: In order to indicate 10, a counter in the lower part on the next upright to the left is put in, i.e. a counter in tens' place is put in.

In illustrating the manipulations which are most important in calculation's, the following symbols are used to facilitate one's understanding of its manipulation.

Push up the counter with the thumb or forefinger.

(In the case of taking off the counter in the upper part, push up the counter with the forefinger. (subtraction)

In the case of putting in the counter in the lower part, push up the counter with the thumb. (addition)

Push down the counter with the forefinger. (means addition in the case of five counter) (means subtraction in the case of one counter) A counter already placed, i.e., a counter which has already been "put in" and which remains either on the top or at the bottom.

(*) A counter newly "put in".

(/) A counter "taken off".

Reading the Abacus of Counting

The counting rods (a kind of numeral) use a combination of horizontal and vertical lines. Each numeral consists of two sets of lines. Each line in the bottom set has a value of 1, and each line in the top set has a value of 5. The value of a numeral is found by adding the values lower of the top and bottom lines. For instance, 7 is represented by (two lower beads with a value of 1 each, plus one upper bead with a value of 5). (see Fig. 4a)

Larger numbers are also processed the same way. Again, keeping in mind the values for the beads and the vertical pegs, the number 97 is represented by 1 bead from Heaven and 2 from Earth on the first peg, and 1 bead from Heaven (representing 50) and 4. beads from Earth (representing 40) moved towards the divider but on the second peg.

Counting to 12

Begin by moving the first bead on the first rod on the right to the beam. This equals one. Push the second bead on the first rod up against the first one. This equals two. Push up the third bead to make three- For four, push up the fourth bead, and for five, push up the fifth bead. Now you have five earth beads from the ones rod against the cross-beam. Because each heaven bead equals five, you can -now push the bottom heaven bead on the same rod down to the beam, and drop all of the earth beads back to the bottom of the frame. This represents five. To make six, push the first earth bead from the first rod back to the beam. Now you have a five heaven bead and a one earth bead, making six. For seven, push up the second earth bead on the same rod; for eight, the third bead on the rod; for nine, the fourth bead on the rod; and for io, the last bead on the rod. Now you have 10. The second rod is the tens rod, so move the first earth bead on the tens rod to the beam, and push the heaven and earth beads on the ones rod back to their clear positions. To make ii, leave the tens bead where it is and slide the first earth bead on the ones rod to the beam. For 12, slide up the second earth bead on the one rod. And so on. For another example, 7+7=14, see Fig. 4b below:

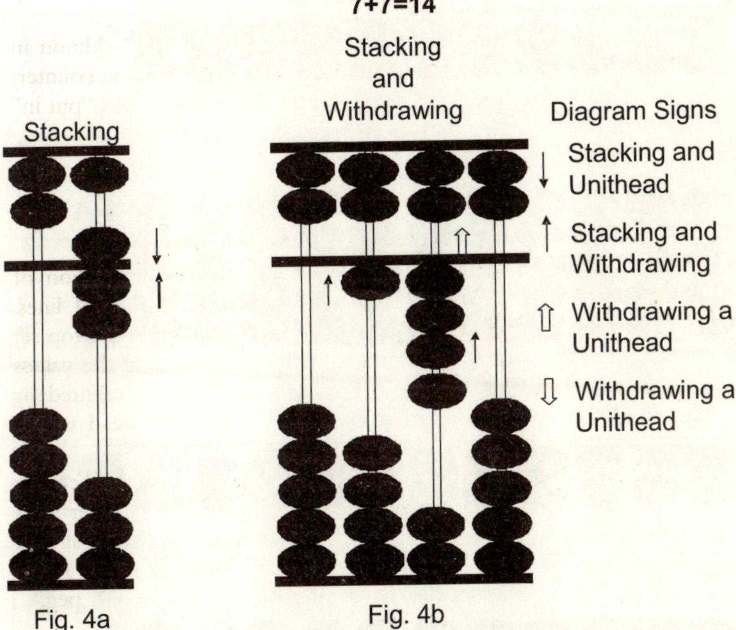

Fig. 4a

Fig. 4b

Thus, any wire in the abacus can be read as a single number, and compared to a Chinese rod numeral, or to a modern Hindu-Arabic numeral. The following Figs. 5(a-l) show the movement of beads to get the results given below each Figure.

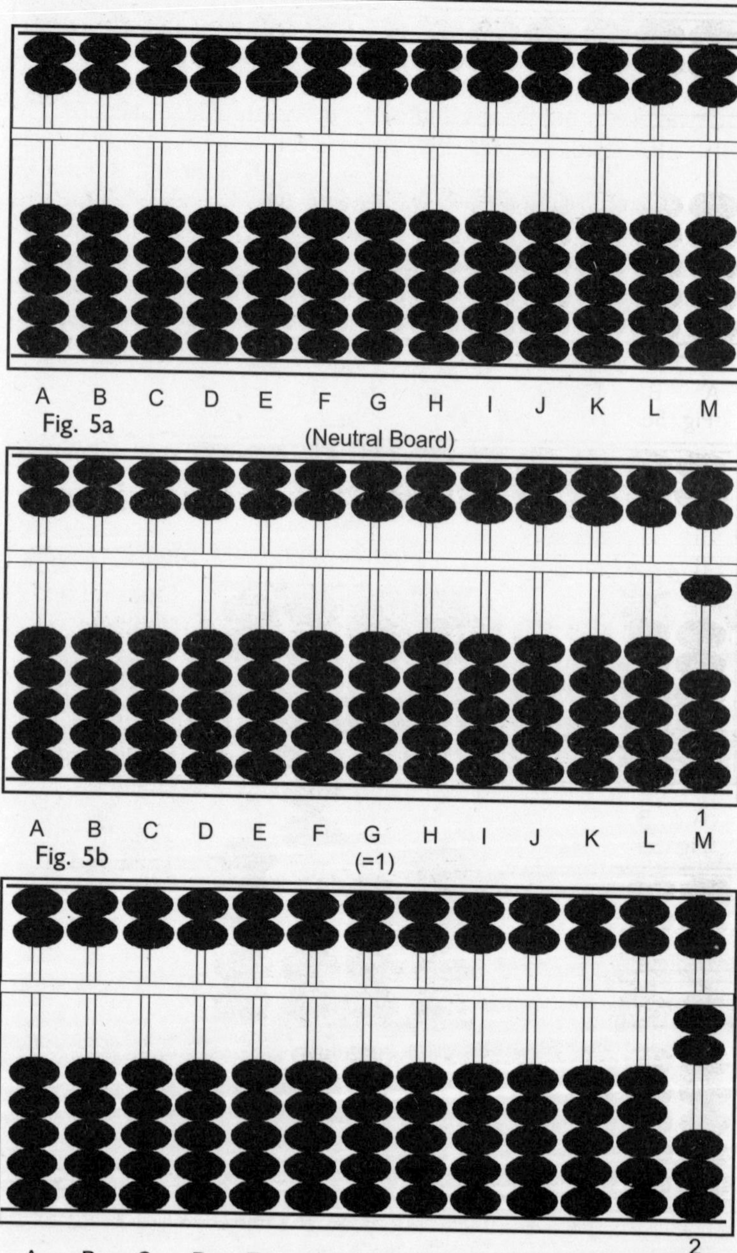

Fig. 5a

Fig. 5b

Fig. 5c

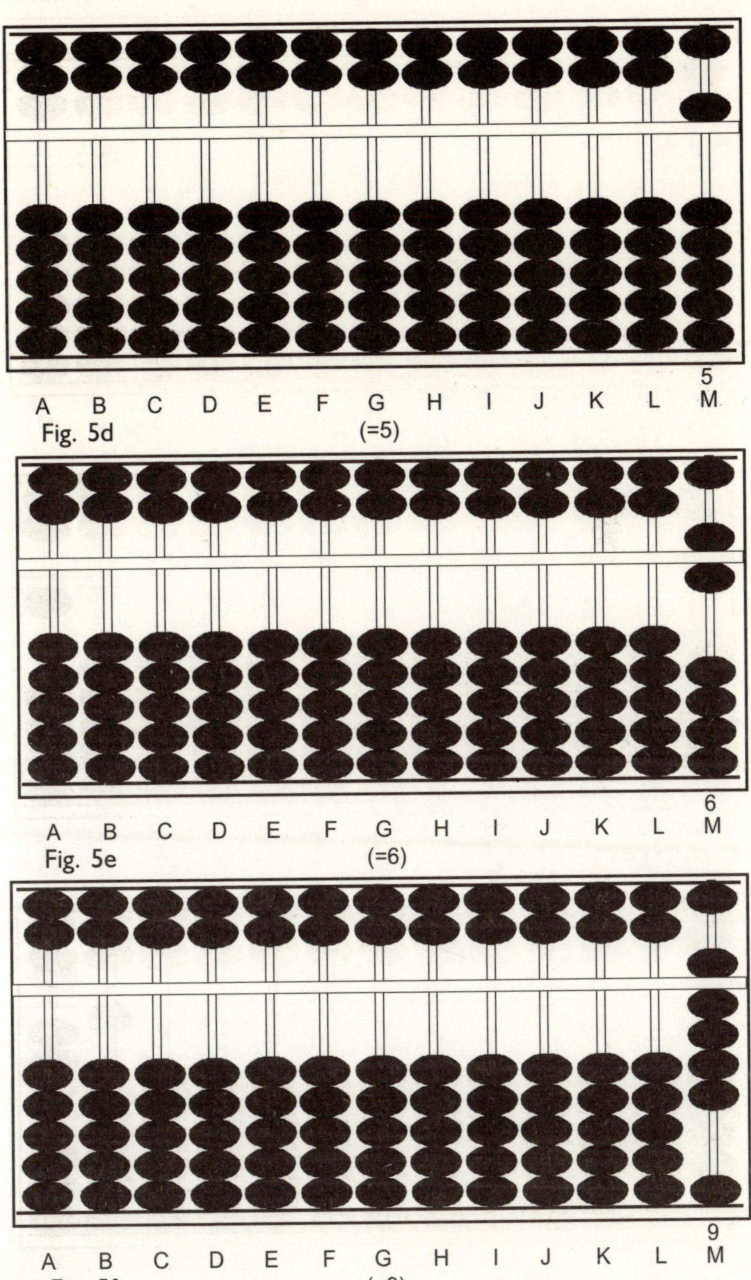

Fig. 5d

Fig. 5e

Fig. 5f

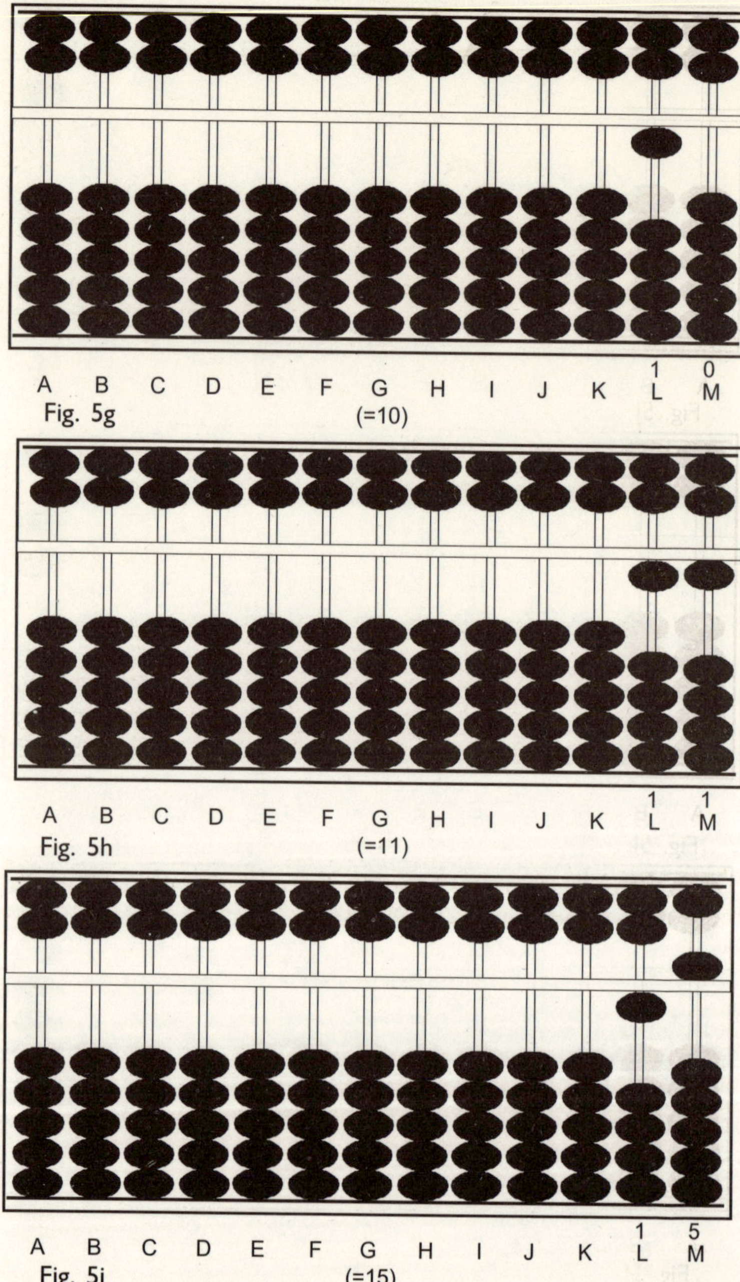

Fig. 5g (=10)

Fig. 5h (=11)

Fig. 5i (=15)

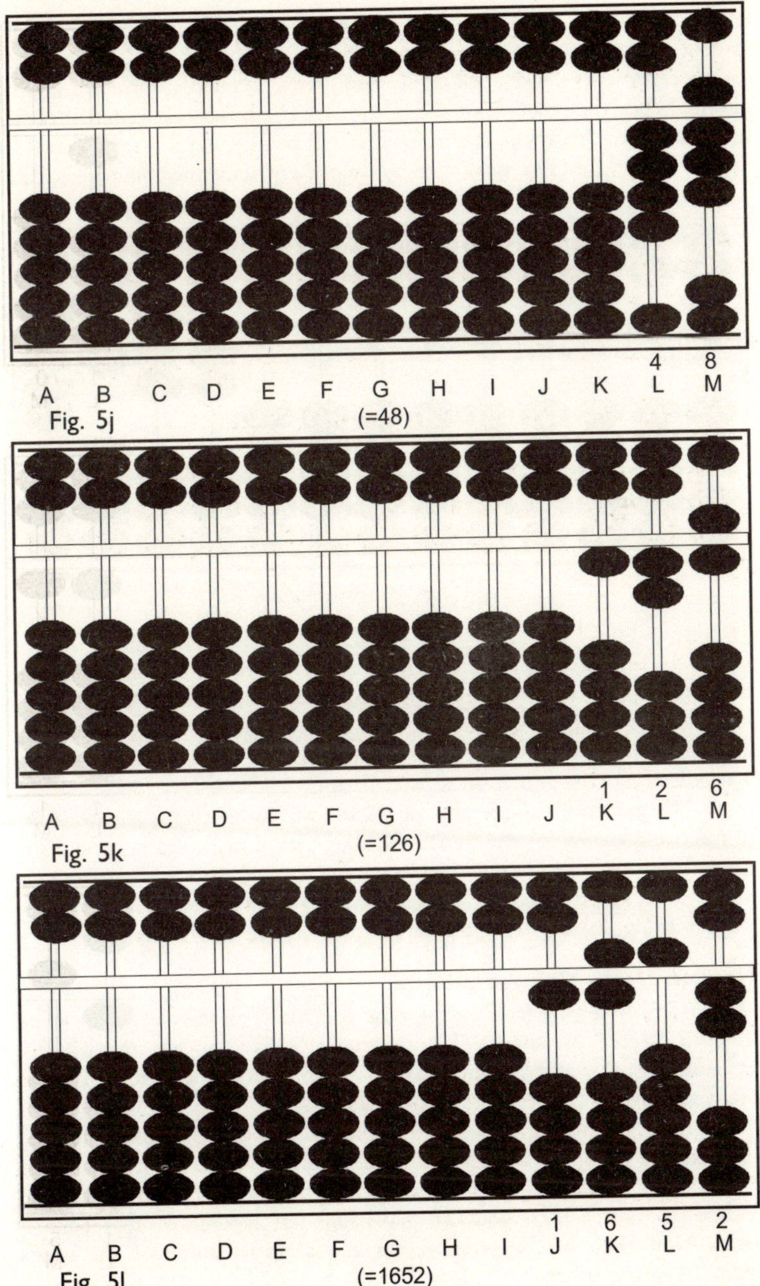

Fig. 5j

Fig. 5k

Fig. 5l

Numbers are represented on an abacus using the same kind of positional system that we are familiar with in the West. Each wire of beads represents a place value, i.e. ones, tens, hundreds, thousands, and so on. While the position of the ones place is arbitrary (in order to allow the abacus to represent decimals), we will put the ones on the farthest wire to the right. The next wire will represent tens, then hundreds, &c. Under this system, 2,718 would be represented as: (see Fig. 6)

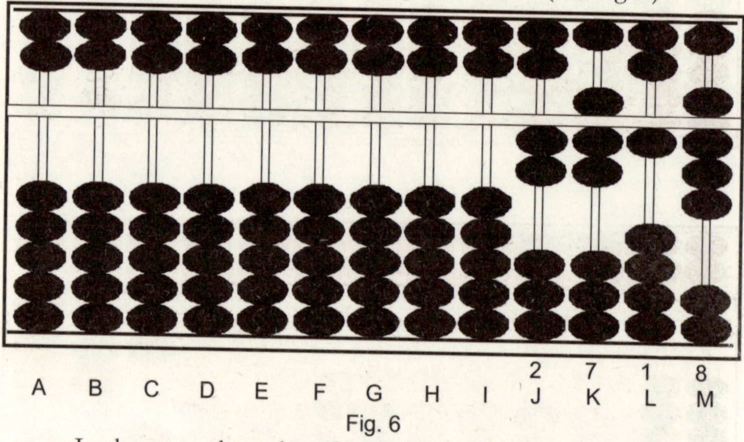

Fig. 6

In the ones place, three beads from the lower de and one bead from the upper deck have been moved to the beam; in the tens place, .one bead from the lower deck has been moved to the beam; and so on. The digits are read from left to right, giving us a value of 2 thousands, 7 hundreds. I ten, and 8 ones (the number 2,718).

Larger numbers are also processed the same way. Again, keeping in mind the values for the beads and the vertical pegs, the number 97 is represented by 1 bead from Heaven and 2 from Earth on the first peg, and i bead from Heaven (representing 50) and 4 beads from Earth (representing 40) moved towards the divider but on the second peg.

How to Use a Chinese Abacus

The first step in knowing how to "read" an abacus is- knowing how to -"read" an abacus. I hope that you have an abacus yourself, because trying examples and playing around with it a little will make learning how to use it much, much easier.

As told above, the abacus consists of 13 columns, each one divided into an upper deck and a lower deck; the lower deck consists of 5 beads per column, the upper deck has 2 beads. per column. I include a little diagram, I hope you can make sense of it: each of represents a bead. Note there is always space for you to move some of the beads away from the others.

Diagram of a Typical Abacus

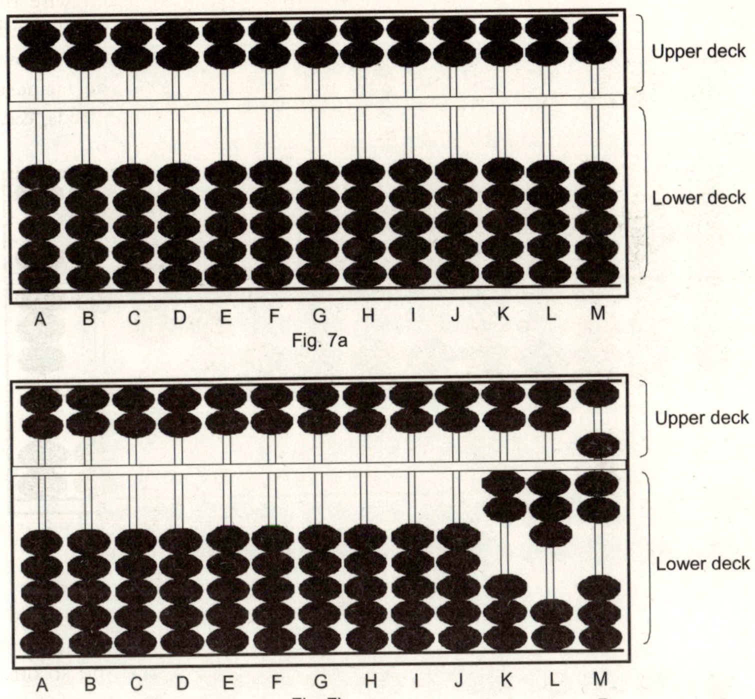

Fig. 7a

Fig. 7b

13 columns (can be named A to M) (see Fig. 7a above). Each column represents a digit, so with 13 columns you can do calculations with numbers up to 10 trillion

The way to express a number is to move up the beads of each column to represent a digit, starting from the right. The value of a bead in the lower deck is 1, the value of a bead in the upper deck is 5. For example, to express the digit 7, you would move up two beads in the lower deck and one bead in the upper deck (see dark digits in column M) so that (1+1)+(5)=7. Make sense?

Knowing that, this is what the number 237 would look like: (in column KLM). (see Fig. 7b above)

You might notice that each column can take on a value from 0-15 and think it's silly because in our number system, each digit only takes on values from 0-9. However, this becomes handy when you need to carry digits that exceed 10, just like you do when doing addition with pen and paper. Also, it allows you to do calculations with other number systems that are not 10-based, if you want to.

Let us now consider an example: 529. It is shown in the Fig. 8b below:

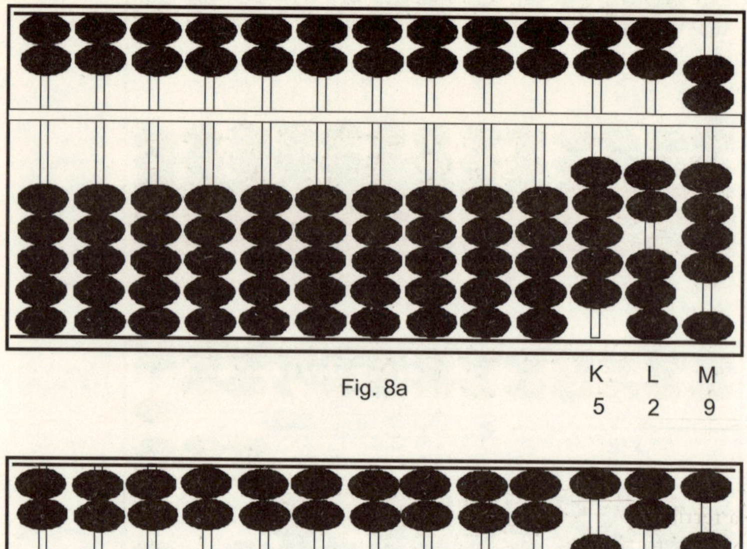

Fig. 8a

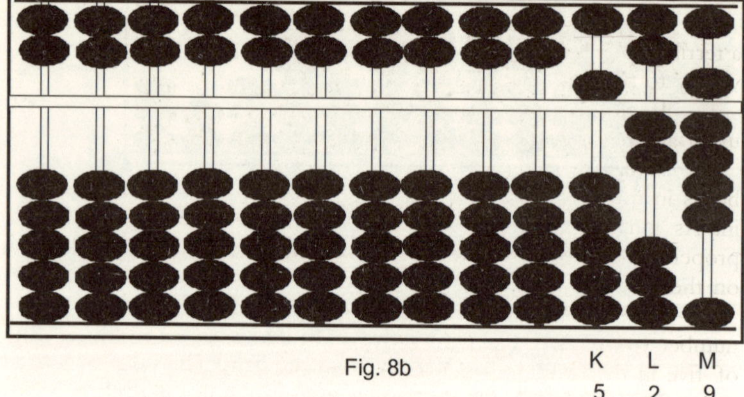

Fig. 8b

Notice that the digit 5 in the third column K (Fig. 8a) which are moved up in lower deck, can alternatively be expressed by moving down one bead in the upper deck of column K. (see Fig. 8b).

Addition

Being able to read numbers off of an abacus is nice, but is not a terribly useful skill by itself. The real power of the abacus comes when we try to perform computations.

If all this starts, making sense, I think you can already see how using an abacus can be useful for simple arithmetic.

Addition on the abacus involves registering the numbers on the beads in the straightforward left-to-right sequence they a written down in. As long as the digits are placed correctly, and the carry's noted properly, the answer to the operation immediately presents itself right on the abacus.

Move beads to the center beam to represent the digits of the number you wish to enter. In each row, the heaven beads have a value of five in that row, and each earth bead has a value of one.

In order to add on the abacus, we need to be able to do a very little bit of mental arithmetic. Specifically, we need to be able to add to either 5 or 10 in one bead.

However, the abacus is only useful if one can "read" it quickly enough and do the simple additions in your head (namely, 1 upper bead and 4 lower beads=1*5+4*1=9), but that really isn't hard to learn.

Basically, we need the facts 1+4=5, 2+3=5, 1+9=10, 2+8=10, 3+7=10, 4+6=10 and 5+5=10. If we can do this, everything else is purely mechanical.

Basic addition works as we would intuitively expect. Given two numbers that we want to add, we register one number on the abacus, then add the second number by moving beads on the appropriate wire.

There are 4 approaches to performing addition, each applied to a particular situation. Each of these techniques is explained in tabular form in the sections that follow.

SIMPLE ADDITION

When addition 6+2, one would move 1 bead from the upper deck down (value=5) and one bead from the lower deck up (value=1); this represents 6. Moving 2 beads from the lower deck (in the same column.) to (value=1*2 beads 2) would complete the operation for addition. The answer is then obtained by reading resultant bead positions sum=8.

		Move bead(s) in the counter	
Given the first number	*To add*	*in the lower deck*	*in the upper deck*
0, 1, 2, 3, 5, 6, 7 or 8	1	+1	
0, 1, 2, 5, 6 or 7	2	+2	
0, 1, 5, or 6	3	+3	
0 or 5	4	+4	
0, 1, 2, 3 or 4	5		+1
0, 1, 2 or 3	6	+1	+1
0, 1 or 2	7	+2	+1
0 or 1	8	+3	+1
0	9	+4	+1

Note: '+' represents the operation of moving the bead counters close to the beam.

Addition

Let us perform the following additions on abacus.

(A) 2+1

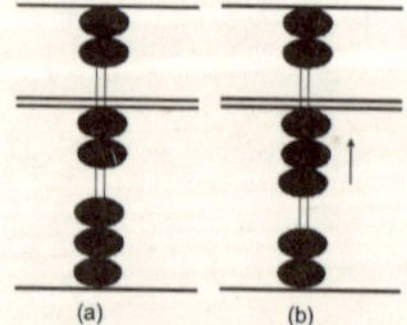

In this case, first put in 2 beads with the thumb towards the beam, then put in 1 with thumb as shown.

Answer: 3

(B) 2+6

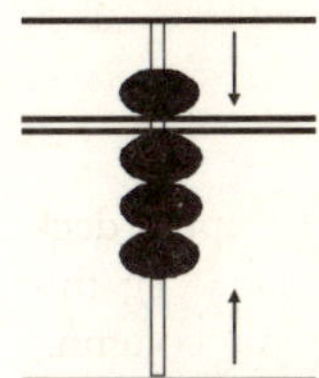

In this case, put in 2 with the thumb first.

Then put in 5 (1 bead in upper part) with the forefinger, and simultaneously put in 1 (bead in lower part) with the thumb as shown in figure.

Answer: 8

Example 1: To add 876,543,210+123,456,789. This addition is described if Figures below:

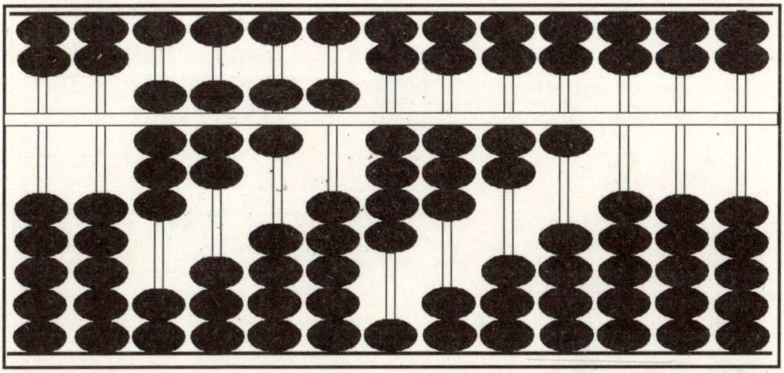

A	B	C	D	E	F	G	H	I	J	K	L	M
		8	7	6	5	4	3	2	1	0		

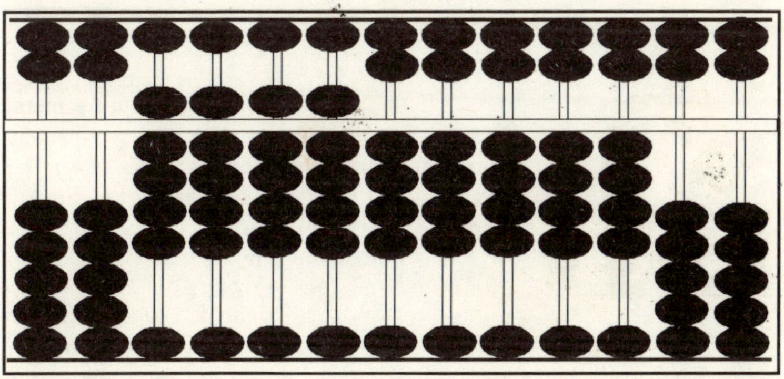

A	B	C	D	E	F	G	H	I	J	K	L	M
		8	7	6	5	4	3	2	1	0		
	+	1	2	3	4	5	6	7	8	9		
		9	9	9	9	9	9	9	9	9		

Combined Adding-up and Taking Off

When the original number registered on a rod is smaller than 5, but will become greater than 5 after the addition is made. The operation requires to move one bead frorn the upper-deck down to the beam and one or more beads from the lower deck removed from the beam.

For example, when 4 is to be added to 3, the operation required is to move one bead counter of the upper deck down to the beam and one bead counter off the original 3, leaving only two bead counters to the pack of the lower deck, which, together with the one bead counter of the upper deck, gives the answer 7.

For example: To add the numbers 7+5, you would first place number (7) on the abacus board on the first peg. Then, since the number will total greater than 10 upon completion, the 1 bead from Heaven on the first peg is put back into place, and 1 bead from Earth on the second vertical peg is brought to the center. This then becomes the representation for 12.

(C) 2+4

2 +4 6

(a) (b)

First, put in 2 with the thumb; Second, to add four put in 5 of the upper part and take off 1, of the lower part with the forefinger (See Figures (a) and (b)..

Answer: 6

(D) 9+7

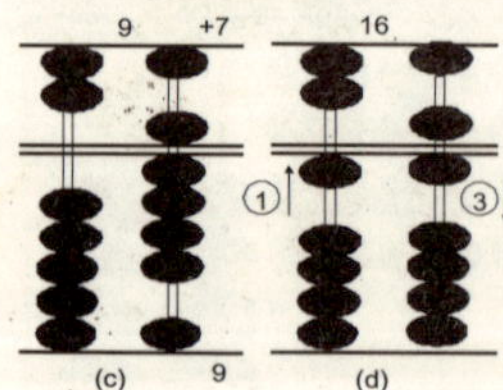

(c) 9 (d)

First, put in 9 with the forefinger and thumb. Then to add 7, take off 3 with the forefinger, and put in 1 in place with the thumb in left hand coloumn beads, i.e., think 7 and 3 are 10. So take off 3 and carry the 1 ten which is put in the tens' the tens' place with thumb (See Figures (c) and (d)..

Answer: 16

When a sum greater than 10 occurs on a certain rod, beads are removed from either or both the upper and lower decks and 1 bead i added to the rod directly to the left. Example: When adding 9 (10–1) to 8, one bead from the lower deck is removed (–1) and one bead from the lower deck on the row directly to the left is added (+10).

		Move bead(s)	
Given the first number	*To add (formula)*	*in the lower deck*	*in the upper deck*
4	1 (+5–4)	–4	+1
4 or 3	2 (+5–3)	–3	+1
4, 3 or 2	3 (+5–2)	–2	+1
4, 3, 2 or 1	4 (+5–1)	–1	+1

Note: '—' represents the operation of 4moving the bead counters off the beam.

Now, let's **add 8+4** on the abacus.

Step 1

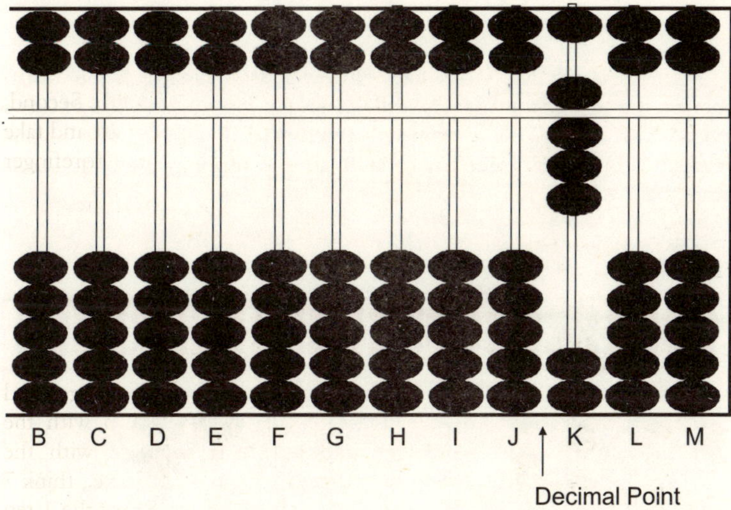

Move the beads on the one's place value bar 'K' directly to the left of the decimal point) to represent the number 8. We use a 5 point bead and three 1 point beads.

Now, we need to add four more. **But** notice, there aren't four more single beads that can be moved, only a 5 point and two 1 point beads.

So, we will **make four** by moving the one 5 point bead down and then removing one of the 1 point bead, (because 5–1=4):

Step 2

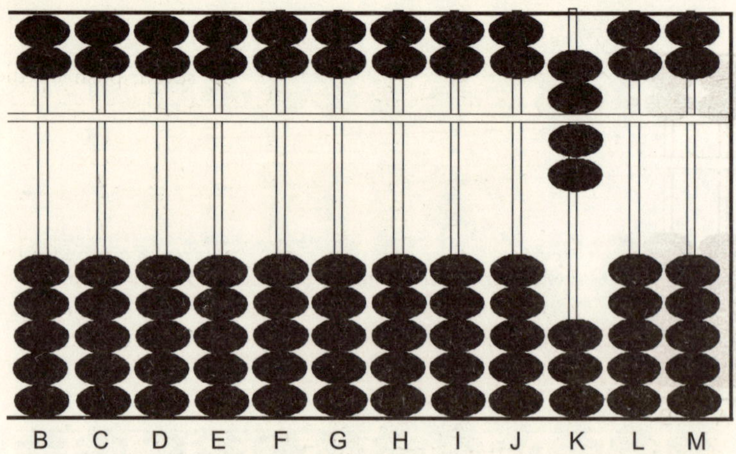

This is what it looks like. However, it is not in proper form, because the two 5 point beads really make ten. So we have ten plus two more which is 12. This is properly represented by moving one 1 point bead up on the tens place bar J value bar and pushing the two 5 point beads back on bar K.

Step 3

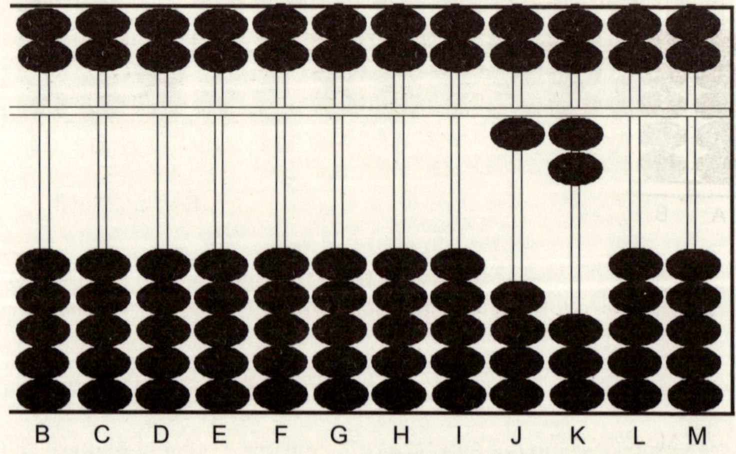

So, here you see the final answer **12**.

Another Example is to add 1,234,444+4,321,432. This addition is described in Figures below.

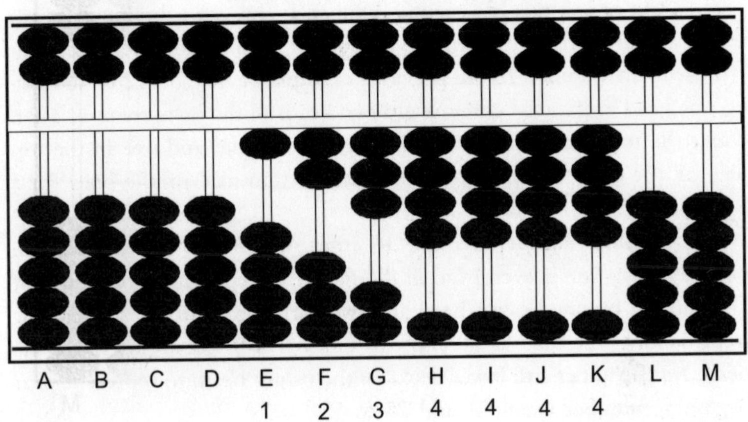

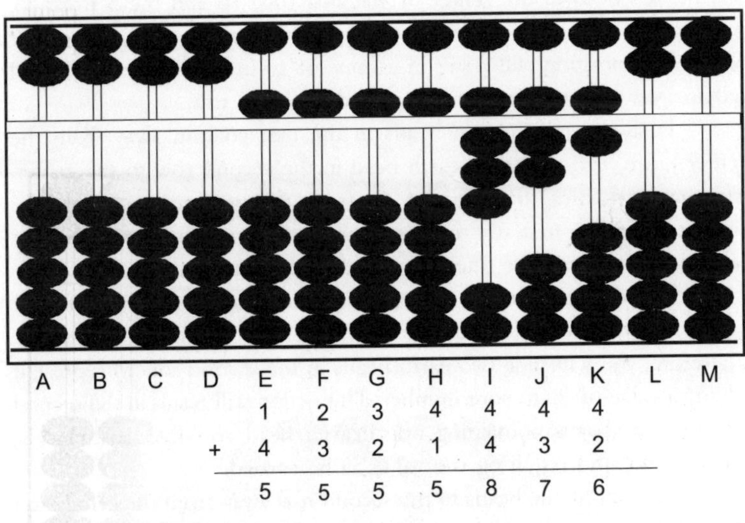

Move beads to represent the number you are using. For instance, if the number is 46 you will move four beads up in the tens column and six in the ones. There are only live beads in the ones column, but since the top portion is five times that, you can move one top bead in the five column and one in the bottom portion for a total of 46.

Add or subtract the second number. If you are subtracting, move the beads down from the original number, if adding, add the beads that represent the number. In the previous example of 46, add 12 by moving up one bead in the tens column and two beads in the ones column. Then count the total beads to get five in the tens column and one in the top half of the ones column and two in the bottom half of the ones for a total of 58.

Consider another example to enter the 75 into the abacus, you need to move one heaven bead in the first row (representing one unit of 5), followed by one heaven bead and two earth beads in the second row (representing one unit of 50 and two units of 10). So, always move the beads on the upper and lower decks to the center beam to represent your beginning number example, add 25 to 75.

Slide the beads in the first row (on the right hand side of the abacus) to add the first digits of each number. In this example, 75 left one heaven bead in the center of the first row. To add the 5 from 25, you need to move one more heaven bead in the first row to the center beam. This position will result in a value of to in the first row, and will require you to carry the number into the second row.

Push the two heaven beads in the first column away from the center beam, and push one earth bead in the second row to the center. This action carries the value of to from the previous step into the correct row. The first row of the abacus will now be empty, and the second row will include one heaven bead and three earth beads.

Side the beads in the second row to add the second digit of your number. In this example, you have the 20 from 25 left to add to your 75. Therefore, you will slide two earth beads in the second row to represent adding a value of 20 to your number. This action will result in the second row of the abacus containing one heaven bead and five earth beads, totaling 100, and requiring the value to be carried.

Push all of the beads in the second row away from the center and slide up one earth bead in the third row. The abacus will now feature empty first and second rows and one earth bead in the third row, resulting in a total of 100.

Example 1: Evaluate 36+75

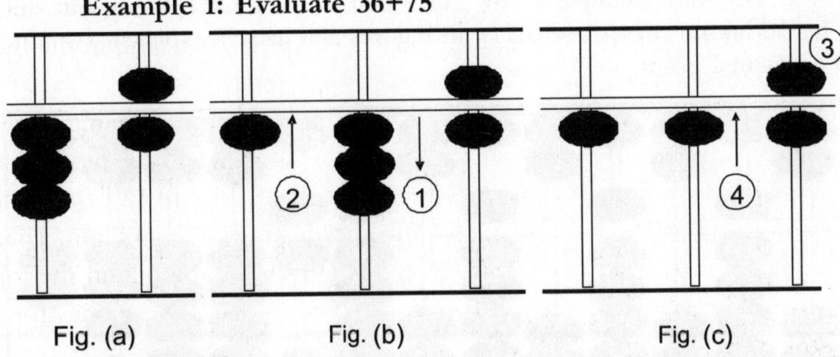

Fig. (a) Fig. (b) Fig. (c)

Answer: 111

To add two place numbers and above, always commence adding from the left to the right.

In this case put in 7 tens i.e., add 7 tens of 75 to 3 tens of 36 in the same method as in example D. Next add 5 ones to 6 ones in the following way.

Take off 5 (a five counter) with the forefinger, and simultaneously put in 1 ten in the tens' place with the thumb as in example D.

Combined Taking-off and Place Advancement

When a sum greater than io occurs on a certain rod, beads are removed from either or both the upper and lower decks and i bead is added to the rod directly to the left. Example: when adding 9 (10–1) to 8, one lead from the lower deck is removed (–1) and one bead from the lower deck on the row directly to the left is added (+10).

Given the first number	*To add (formula)*	*Move bead(s)*		
		In the lower deck	*In the upper deck*	*Lower deck, adjacent (left) column*
9	1(–9+10)	–4	–1	+1
8 or 9	2(–8+10)	–3	–1	+1
7, 8 or 9	3(–7+10)	–2	+1	+1
6, 7, 8 or 9	4(–6+10)	–1	–1	+1
5, 6, 7, 8 or 9	5(–5+10)		–1	+1
4 or 9	6(–4+10)	–4		+1
3, 4, 8 or 9	7(–3+10)	–3		+1
2, 3, 4, 7, 8 or 9	8(–2+10)	–2		+1
1, 2, 3, 4, 5, 6, 7, 8 or 9	9(–1+10)	–1		+1

Another example is to add 9,090,909,644.32+1,020,304,567.89. This addition is described below in Figures and us self explanatory. Note the decimal point.

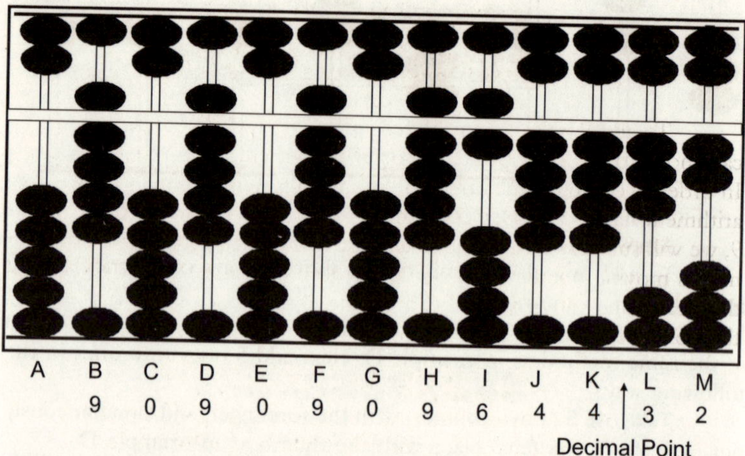

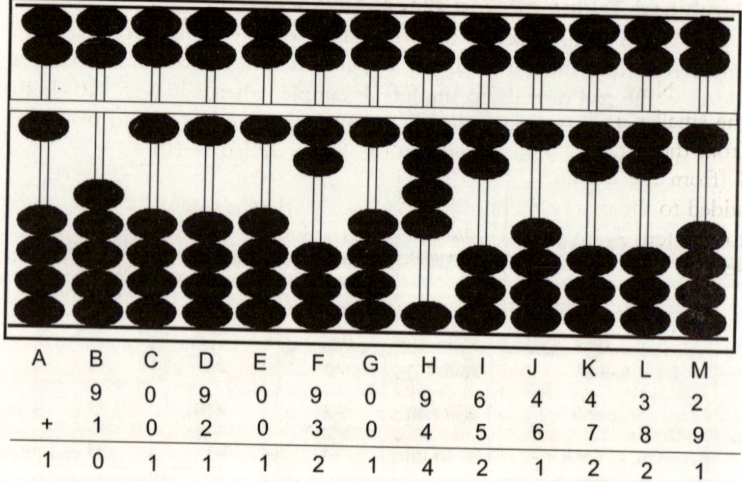

That's all well and good, but what do we do if there are not enough beads on a particular wire? Just like in elementary school, we are going to have to carry something. Consider the problem 18+9 First, register 18 on the abacus Figure (a):

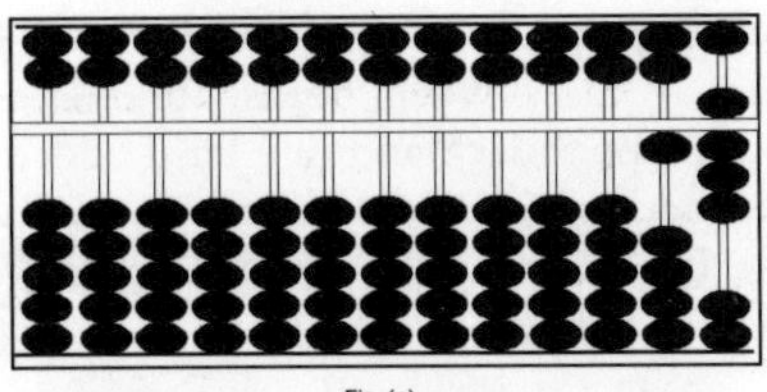

Fig. (a)

To add 9, we start in the tens place. As there is nothing there, we can move to the ones place. Unfortunately, there are not enough beads. In order to complete the operation, we need to remember another of our arithmetic facts: 1+9=10 This implies that 9=10–1. So in order to add 9, we will subtract 1 from the ones wire, and add 10 to the tens wire. The means moving one bead on the lower deck of the ones wire away from the beam, and moving one bead on the lower deck of the tens wire to the beam. The result, 27 (Fig. (b), can now be read from the abacus.

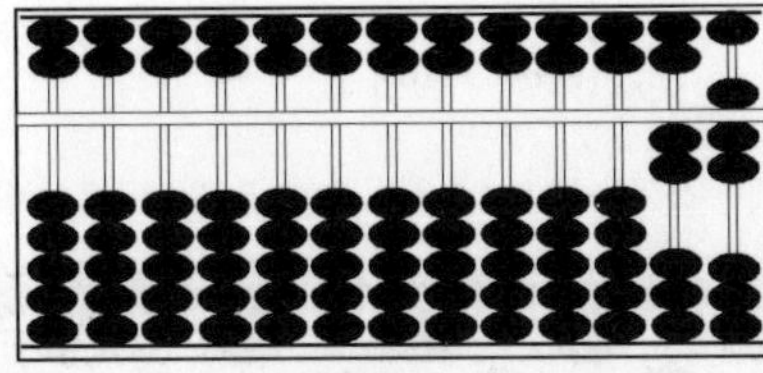

Fig. (b)

Now, when there is one or more double digit numbers to add, put the smaller of the two on the abacus first, the put in the larger number from the left, so 8+12=20 will have 8 on the unit rod first, then the digit I (from the number 12) is placed into the tens column. The digit 2 is added to the units column, but we note it gives 10, so we carry none over to the tens column and clear the units column. (i.e., 2=10–8, so add a second bead to the tens rod, and then take away 8 from the units column)

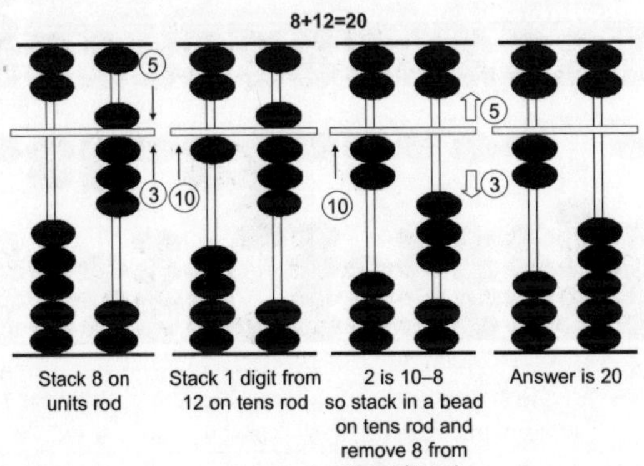

Combined Adding-up, Taking-off and Place Advancement

There are 4 cases when beads are added to the lower-deck, removed from the upper-deck and one bead added to the adjacent rod. Example: When adding 6 (+1-5+10) to 7, one bead is added to the lower-deck, one bead removed from the upper-deck and one bead is added to the left rod (lower-deck).

Given the first number	*To add (formula)*	*Move bead(s)*		
		In the lower deck	*In the upper deck*	*Lower deck, adjacent (left) column*
5, 6,7 or 8	6(+1–5+10)	+1	–1	+1
5, 6 or 7	7(+2–5+10)	+2	–1	+1
5 or 6	8(+3–5+10)	+3	–1	+1
5	9(+4–5+10)	+4	–1	+1

Example 2: Evaluate 806,060,505+607,080,909

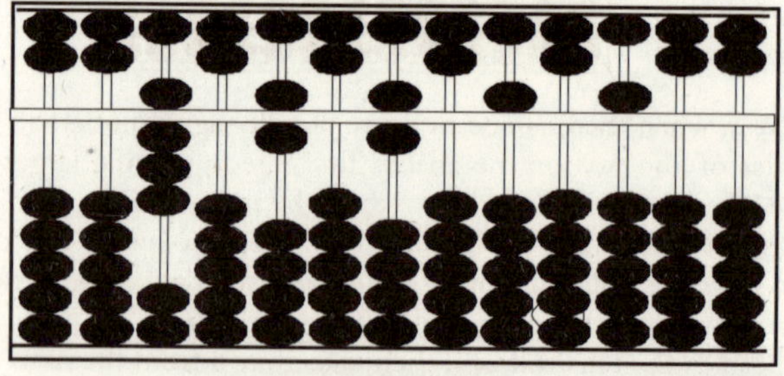

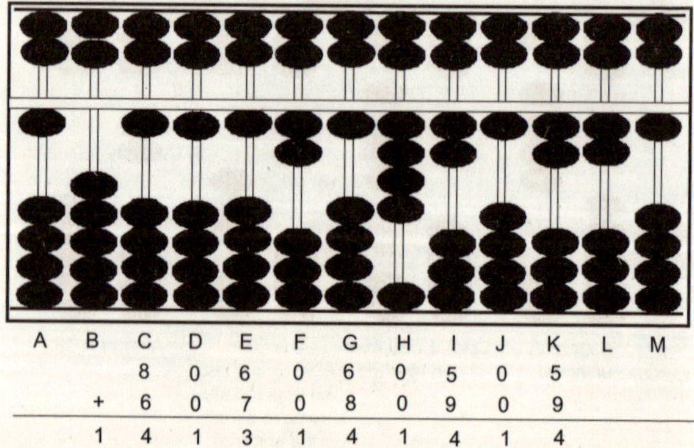

A	B	C	D	E	F	G	H	I	J	K	L	M
		8	0	6	0	6	0	5	0	5		
	+	6	0	7	0	8	0	9	0	9		
	1	4	1	3	1	4	1	4	1	4		

It has been found to be a much simpler and faster process to make additions on an abacus than to do them on paper, especially when there are many numbers to be added together.

The actual procedures can be shown as follows:

Original number

+) 1st addition
1st sum

+) 2nd addition
2nd sum

+) 3rd addition
3rd sum

+) 4th addition
4th sum

=====
=====
=====

(n–1)th addition

+) nth addition
nth sum (Final result)

Below is shown how to add 7 to 7 and give the answer 14 above.

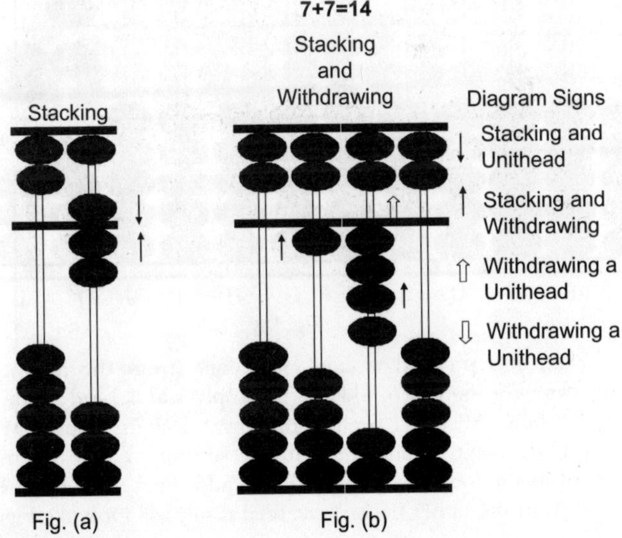

Fig. (a) Fig. (b)

Firstly, seven is stacked by moving a top bead and two unit bead&: to the crossbar (Figure (a)).

Next we have to think differently about the seven we are about to add to the original number seven.

Seven can be thought of as 10 minus 3. i.e. 7=10–3
Also, 3 can be thought of as 5 minus 2. i.e. 3=5–2
By going through two indirect routes, we arrive at the answer.
i.e. 7=10–(5–2)=10–5+2

That is, put a bead in the tens column, remove a top bead (5) and then add two unit beads. Alternatively, the net result of this is that, 7=12–5, so we stack in 12, and remove five (Figure (b)).

Due to the constraints of the abacus' design, doing this route is by far the most natural way to the answer. It is a basic Chinese abacus manipulation technique, which we call SUBSTITUTION. It is modified later when we see how it is used in subtraction.

Obviously if the two numbers add up to less than 10. 50 it tits on a rod, substitution is clearly not used.

Example 3: To evaluate 625+536.

Now, to do something like 625+536, first put down the number 625, in columns KLM as shown in the Figure (a).

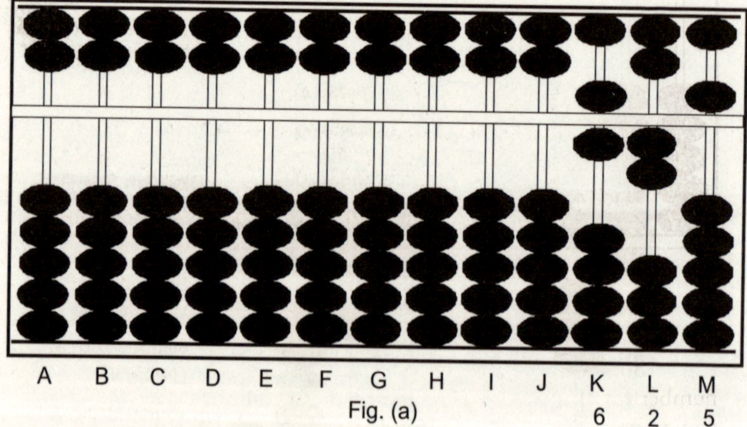

Fig. (a)

Then you proceed to add each digit (from the number 536), starting from the right. To add the 6, simply add 1 bead to the upper deck (in column M) and 1 bead to the lower deck (in column M) (since 1*5+ 1*1=6). Next, to add the 3 to the second column just move up 3 beads from the lower deck (1*3=3). To add the 5 (in column K), add one bead from the upper deck. Your final result will look like Figure (b):

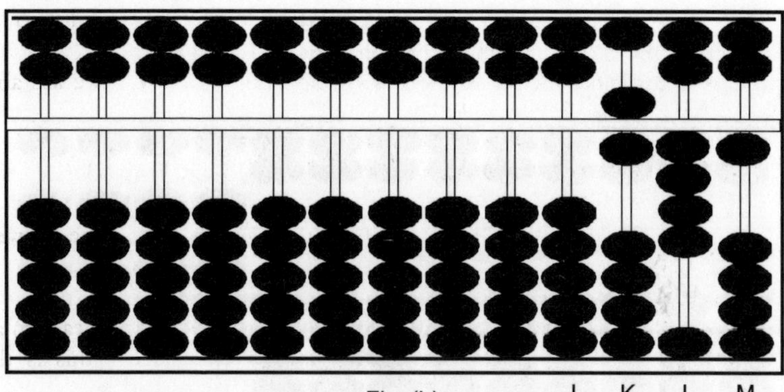

Fig. (b)

Notice that whenever you have 2 beads up in the upper deck (as shown in column K) you can move them down and add one from the lower deck in the next column J (on left of K). If you don't, you will have numbers greater than 10 in a column, which is confusing.

This simplifies to:

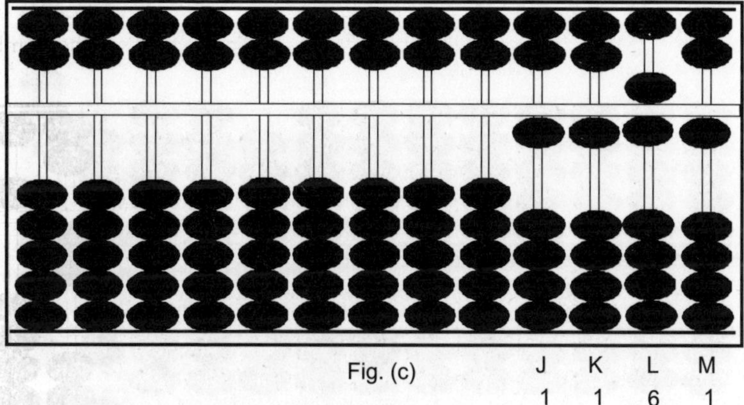

Fig. (c)

It looks tedious, but with practice it becomes much easier.

So what the abacus does for you is that it "holds" the first number for you while you're adding the second one digit by digit. When you want to do addition or subtraction problems, all you're doing is shifting beads up and down each column (i.e. digit) at a time, so that even 23509725–9438558 isn't too difficult to do.

There are lots more "tricks" that one should know to be able to do more addition and subtraction problems, but they are hard to explain without showing. But just for your information, calculations such as multiplication, division, and even taking the square/cube root are all possible, but much more difficult.

As a final example of addition, consider 9,999+77. Again, start by registering the larger number 9999 on rods JKLM (Fig. (a)). The rest process is explained below in Figures (b) and (c).

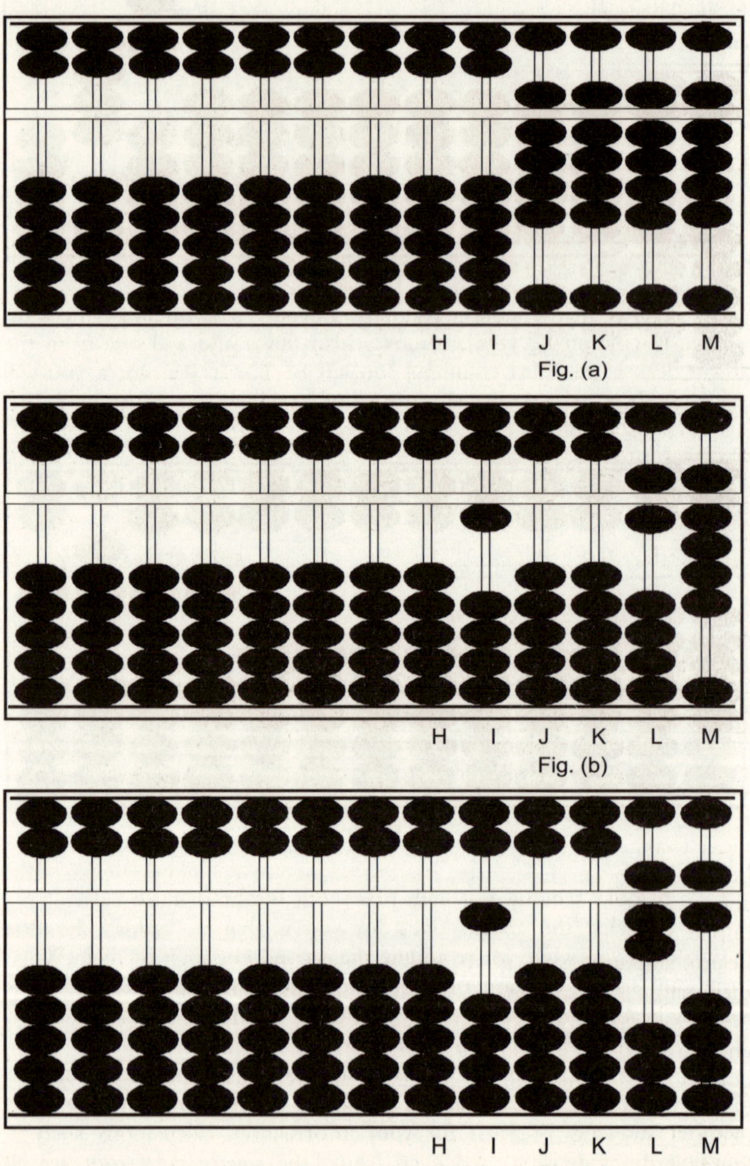

Fig. (a)

Fig. (b)

Fig. (c)

Example 4: To add: **8231+458.**

First, we clear the abacus and register the larger number on the abacus (the beads theirself indicate which ones have moved).

Then, starting on the left and moving to the right, we add the second number:

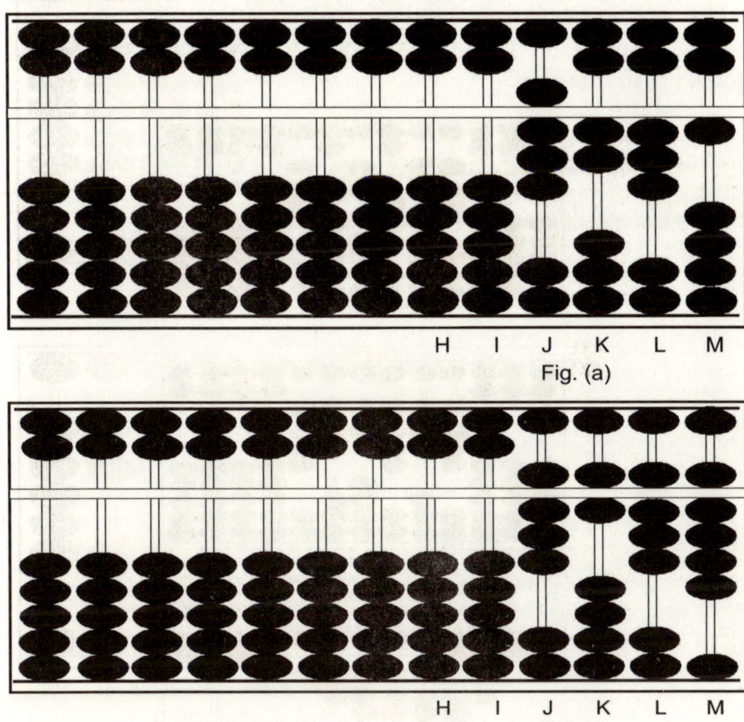

Fig. (a)

Fig. (b)

Starting in the thousands place, there is nothing to add. Then we move to the hundreds place, where we want to add 4. There are not four beads in the lower deck, so we need to use one of the addition facts from above: 1+4=5. Rearranging this slightly, we find that 4=5–1, which implies that we need to add 5 to the upper deck (move one bead to the beam), and subtract 1 from the lower deck (move one bead away from the beam). In the tens place L, we move one bead on the upper deck Next, we need to add 5 to the tens place. Unfortunately, there are not enough beads to do this. So, recalling that 3+7=10⇒7=10–3, we are going to try to add 10 in the hundreds place K (one bead from the lower deck), and subtract 3 from the tens place (three beads from the lower deck). The tens place is easy, but we don't have enough beads in the hundreds place to do what we want. ' What do we do?

We reset that wire, and move to the next wire. On the next wire, we run into the same problem. There are not enough beads on the wire to add one. So, again, we reset the wire and move to the next wire. Finally, we can register our single bead. The result looks like.

Finally, we finish the operation by adding 7 to the ones place M.

The result, 10,076, can be read from the abacus.

Example 5: To add 753,607+509,087+777,179 look at the following four Figures.

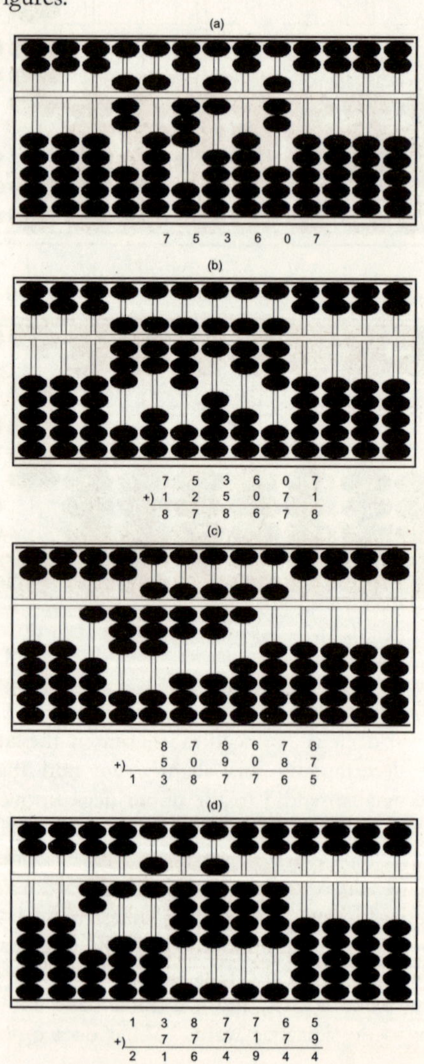

Let's try an addition problem with decimals.

Example 6: Find 1.75+3.32=

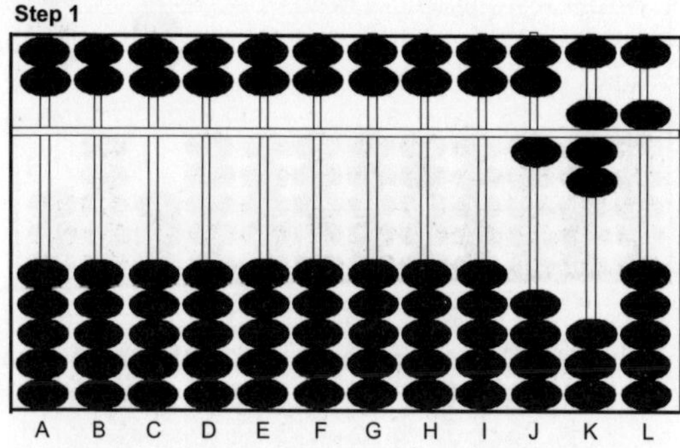

Record 1.75 on the abacus.

Now start adding the 3.32 from the hundredths place value first.

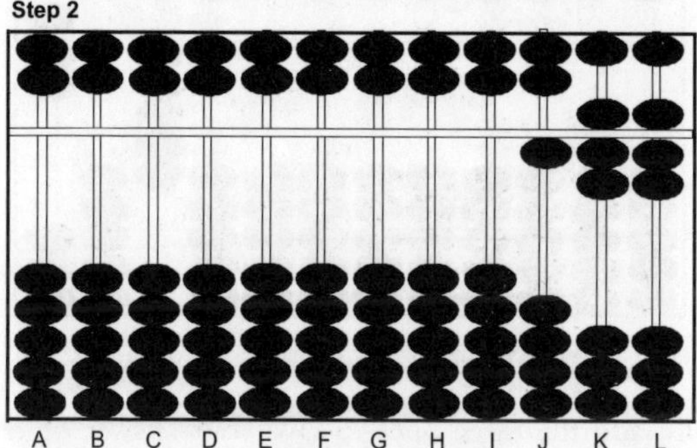

Add two hundredths by moving two 1 point beads up.

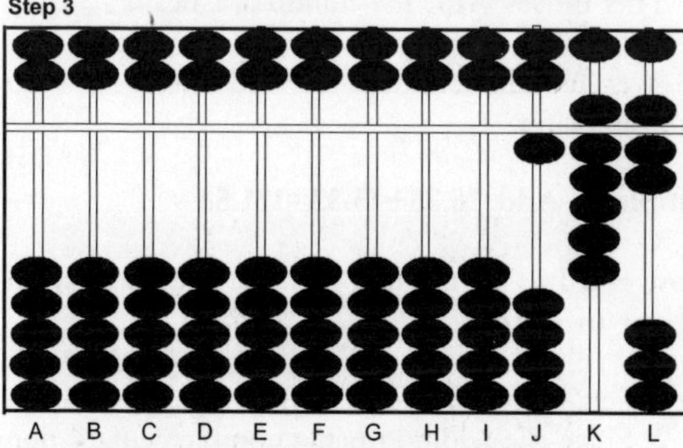

Now move to the hundredths place value bar and add three hundredths by pushing up the remaining three 1 point beads.

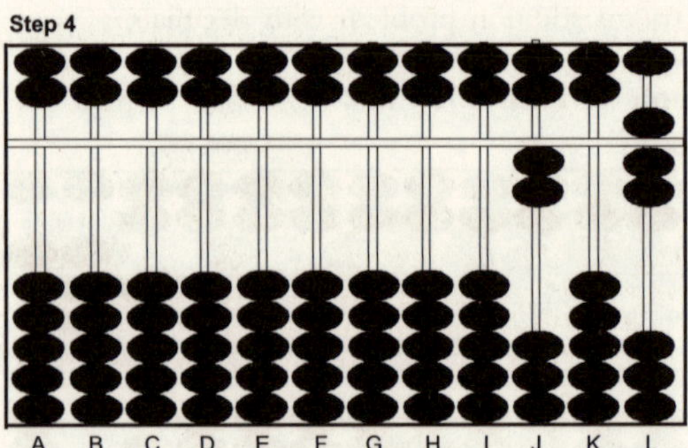

Just push up another 1 point bead on the one's place value bar and then move the other beads on the tenth's value bar back.

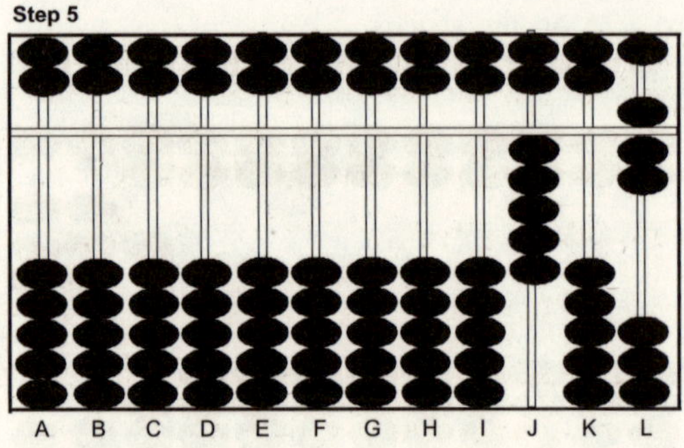

Finally, add the three ones to give your answer of 5.07.

Note: This moves ALL the hundredth beads to the record bar.

This gives us ten hundredths which equals one. So we fix the representation as below.

Example 7: Add 76.25+45.33=121.58

Choose rod I to be the unit rod. Set the decimal on the vernier just to the right of rod I.

Step 1: Set 76.25 onto rods HIJK. (Fig. 1)
Step 2: Solve the addition using complementary numbers where necessary just as you would on a normal suanpan. This leaves the answer 121.58 on rods GHIJK. (Fig. 2)

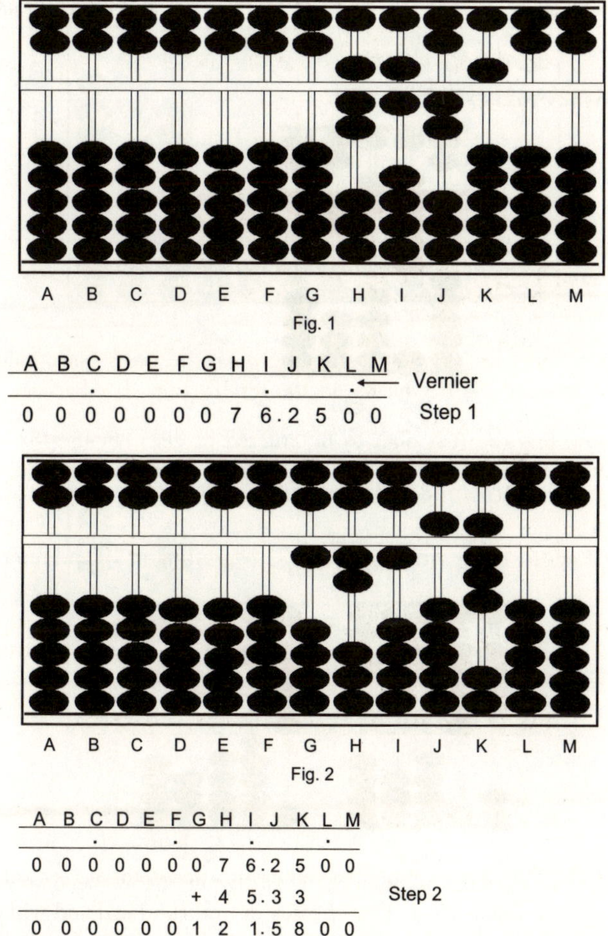

Fig. 1

Fig. 2

Larger Numbers

You will have by now got a good grasp of the basic manipulation techniques for addition. For numbers .2 with a longer string of digits. The processes are just the same, starting from the left and proceeding to the right.

Example 8: To add 86452+9621=?????

As usual if there is a smaller number to be added, then this should be the first number to be stacked into the abacus. In our example here, we will stack in 9621 first rod JKLM (Fig. (a)) and then add 86452, i.e.,

86452+9621=?????

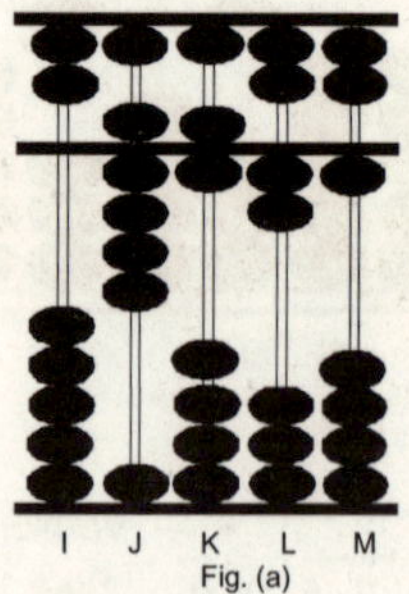

Fig. (a)

Now, 8 is the first digit to be entered in 9621+86452=8....

Other than noting that 8 is an order of magnitude greater, so it is stacked one rod I to the left of the highest place of 9621. There are no surprises here (Fig. (b))..

6 is added to the first digit of 9621 by putting in 10 and removing 4 (95....)

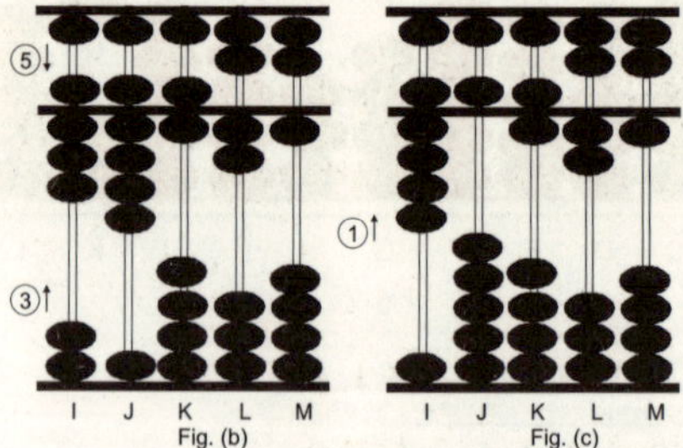

Fig. (b) Fig. (c)

Here we note that to enter the next digit in 86452. (i.e. 6), we have to think of 6 us being 10–4. That is stack in one 10's bead and withdraw 4 unit beads (Fig. (c)).

4 is the next digit and is to be entered in rod K by 10–6 (Fig. (d)).

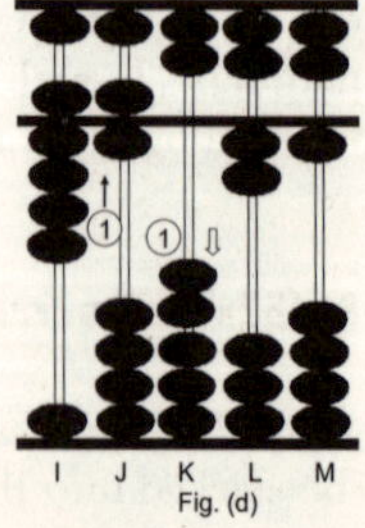

Fig. (d)

i. e. the next digit in 86452, is 4. We have to think of this as being 10–6. That is, stack in one 10's bead and withdraw a topbead worth 5 and a unitbeads worth 1.

The next digit 5 in ten's place is added by moving down a top bead (Fig. (e)).

Fig. (e)

The last digit in 86452, is 2, only requires us to stack two unit beads towards the crossbar (Fig. (f)).

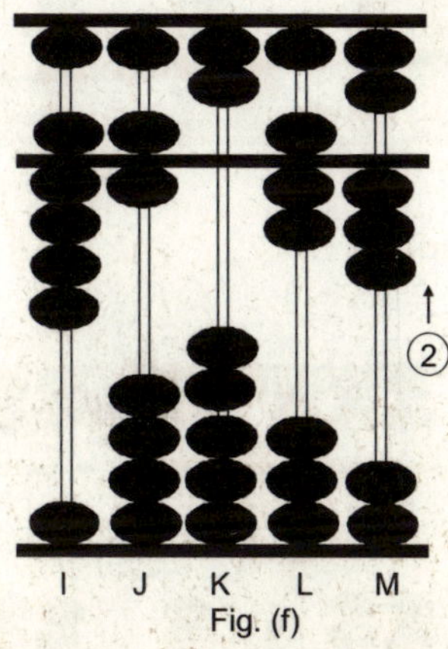

Fig. (f)

Our answer is 96073 for 9621+86452

Some More Examples

Addition on the abacus is straight forward: simply add the numbers in the left to right sequence. (One could add from right to left, the result is the same.)

Example 9: To add 146+52

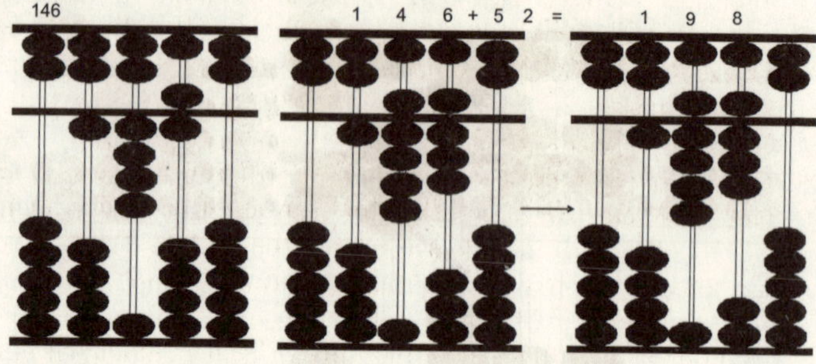

Example 10: To add 42+33

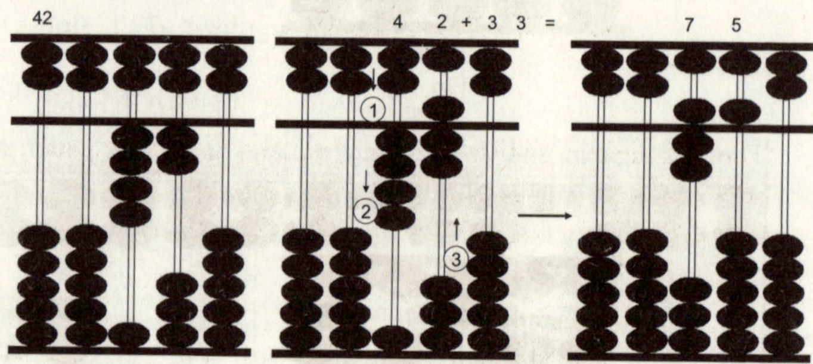

Example 11: To add 28+16

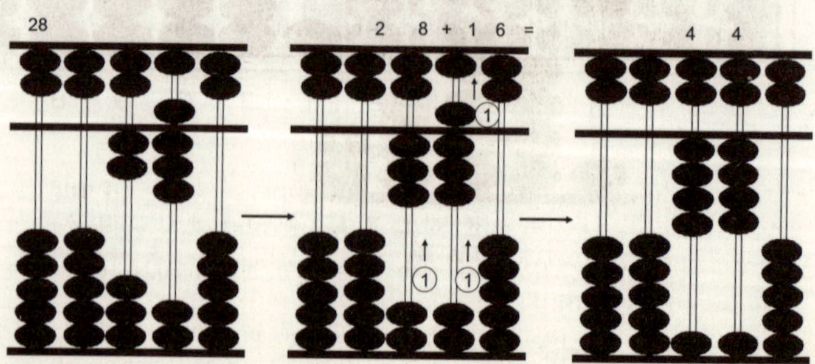

Example 12: To add 378+1659

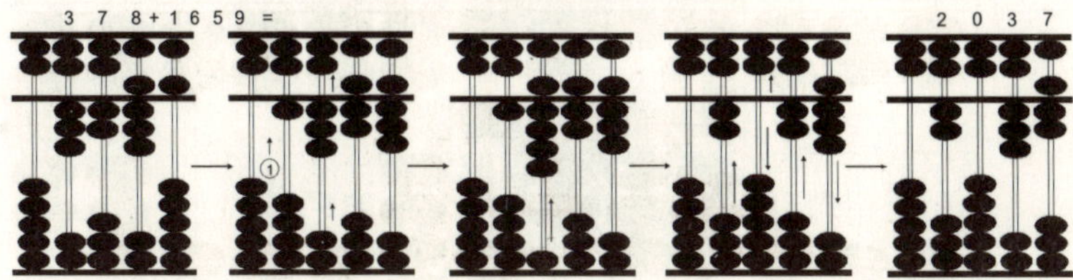

Note when adding we are not simply adding beads but sometimes removing beads. Its mathematical expression, these operations are as follows:

+2=+5–3
+2=+10–8
+6=+10–4
+6=+1–5+10

2

Subtraction

To subtract one number to another, registered the minuend on the abacus; then proceed to subtract place by place (rod by rod, starting from the left, i.e. it involves working from the left hand side to the right hand side of the abacus, involving the use of a slightly different substution method. The beads are removed from either or both the lower or upper decks. The final bead-positions represent the answer.

For all cases, the techniques involved in making subtractions are as follows.

SIMPLE TAKING-OFF

This is achieved by simply taking off one or more beads from the lower deck, or sometimes both. Example; when subtracting 7 (represented by (–5–2=–7) from 9 (which is formed by one upper bead and four lower beads) remove 1 bead from the upper-deck (–5) and 2 beads from the lower deck (–2). The remaining 2 beads represent the result.

The two bead counters left in the lower deck given the remainder 2. (see on next page)

Note: The "–" symbol in the Move bead(s) columns represents moving the bead(s) away from the middle beam.

So, numbers less than 10 do not require the substitution method since they can be handled on the rod itself. So, 9–6 would just meant that a topbead and a unitbead would be removed from the abacus representation of 9, leaving three unitbeads.

Given the first number	To Sub-tract	Use For-mula	Move bead(s)	
			In the lower deck	In the upper deck
0, 1, 2, 3, 5, 6, 7, 8, 9	1	–1	–1	
2, 3, 4, 7, 8, 9	2	–2	– 2	
3, 4 or 8, 9	3	–3	– 3	
4 or 9	4	–4	– 4	
5, 6, 7, 8, 9	5	–5		– 1
6, 7, 8, 9	6	–6	– 1	– 1
7, 8 or 9	7	–7	– 2	– 1
8 or 9	8	–8	– 3	– 1
9	9	–9	– 4	– 1

When subtracting, it is usual to subtract the smaller number from the larger. Since there is no sign to represent negative numbers on an abacus, the operator will have to know if the answer is negative or not by him/herself:

Example 1: Evaluate 10–3

In this case there are no ones. Consequently 1 ten is borrowed by taking off ten with the forefinger and 3 ones are subtracted from it by putting in 7 ones, the remainder, with the forefinger and thumb.

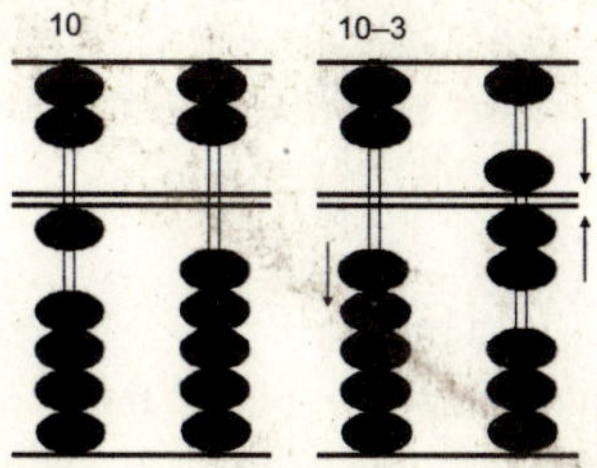

Answer 7

Example 2: Evaluate 12–6

In the same thee are not enough) ones to subtract from. Consequently, as in example above 1 ten must be borrowed. With the forefinger 1 ten is taken off and the remainder 4 ones are put in by putting in a five counter arid taking off a one counter in the ones' place.

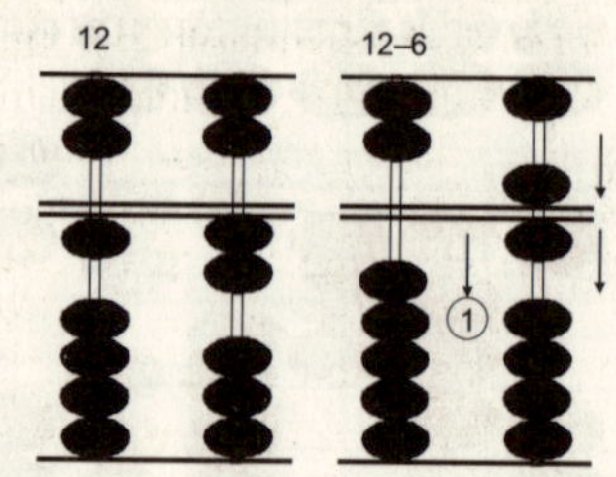

Answer 6

Example 3: Evaluate 12–8=??

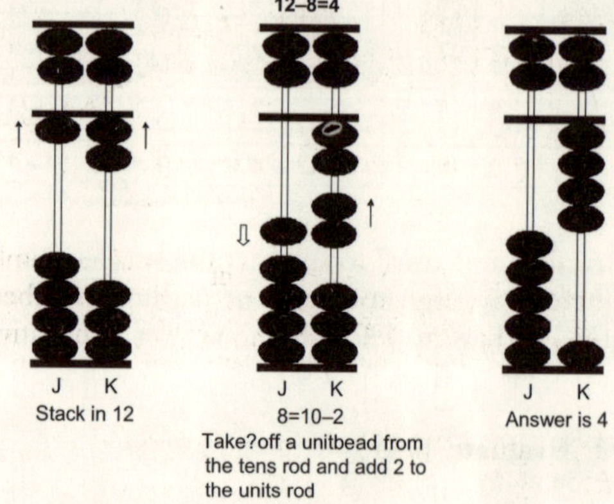

First put 12 on the abacus as shown in Figures.

From our substitution method, we note that 8= 10–2, if we place this into 12- 8 we get,

12–(10–2)=12–10+2 (See Figure)

This means, we take off a uriitbead in the tens column, and add two unitbead into the units column.

Answer: 4

Example 4: Evaluate 100–58

In subtraction involving two place, three place numbers and above, always commence from the left to the right as in the case of addition. In this case subtract 5 tens of the number 58 from the tens. However, as there are no tens, 1 hundred is borrowed in a same manner as in an example above. Next, in subtraction 8 ones, as there are no ones left, 1 ten must be borrowed. 8 is subtracted and 2 ones, the remainder, are put in by the same method as of example (10–3).

Answer: 42

Example 5: Evaluate 999,999,999–123,456,789.

This subtraction is described in Figures (a) and (b) below.

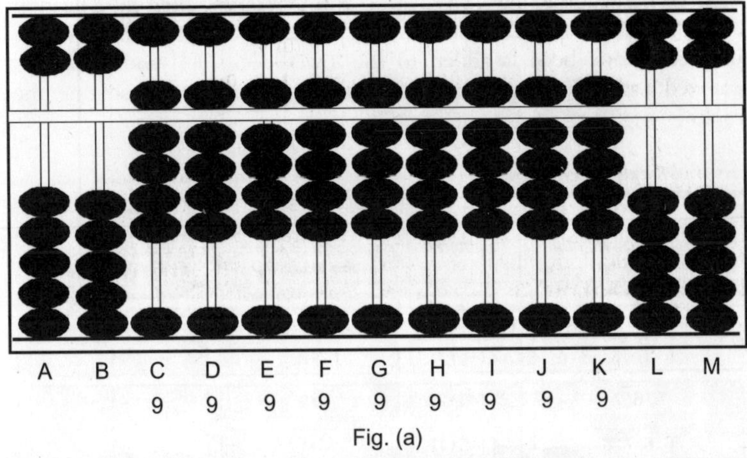

Fig. (a)

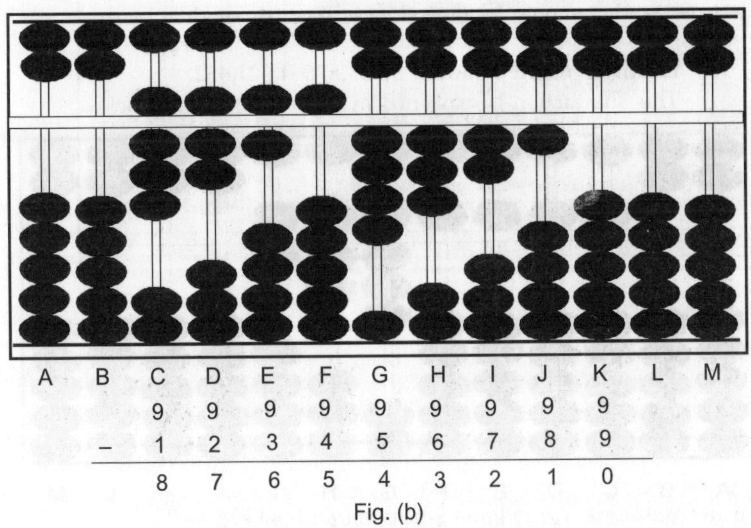

Fig. (b)

Combined Adding-up and Taking-off

When the number of beads in the lower deck is less than the subtracter (the number being subtracted), one or more beads are added in the lower deck and 1 bead is removed from the upper-deck.

When subtracting (+1–5 =–4) from 7 (represented by 1 bead in the upper-deck and 2 beads in the lower deck (less than 4, the subtracter), one bead is added to the lower deck (+1) and 1 bead is removed from the upper-deck (–1) leaving 3 beads, representing the result.

Given the first number	*To subtract*	*Move bead(s)*	
		In the lower deck	*In the upper deck*
5	1 (–5+4)	+4	–1
5, 6	2 (–5+3)	+3	–1
5, 6, 7	3 (–5+2)	+2	–1
5, 6, 7, 8	4 (–5+1)	+1	–1

Note: The "+" symbol in the Move bead(s) columns represents moving the bead(s) towards the middle beam; the "–" symbol indicates that the bead(s) should be moved away from the middle beam.

Example 6: To evaluate 5,555,876–4,321,452.

This subtraction is explained in Figures (a) and (b) below.

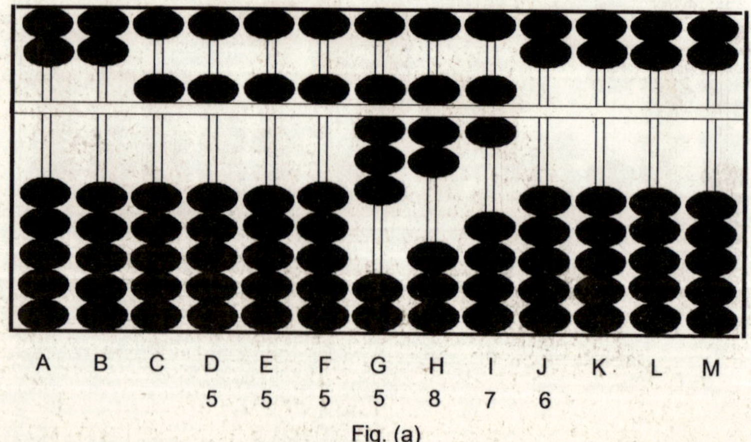

Fig. (a)

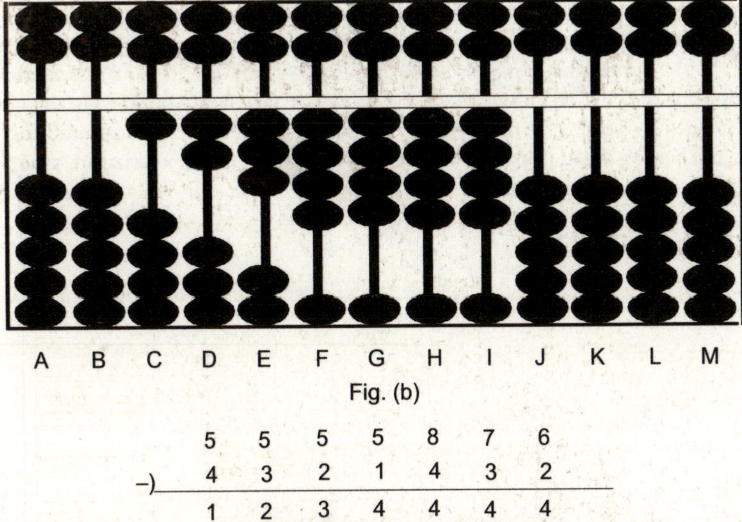

Fig. (b)

	5	5	5	5	8	7	6
–)	4	3	2	1	4	3	2
	1	2	3	4	4	4	4

Subtracting Double Digit Numbers and Higher

The above principle is applied to the above, but we take it a digit at a time, starting again from the left.

Example 7: Find 52–39=???

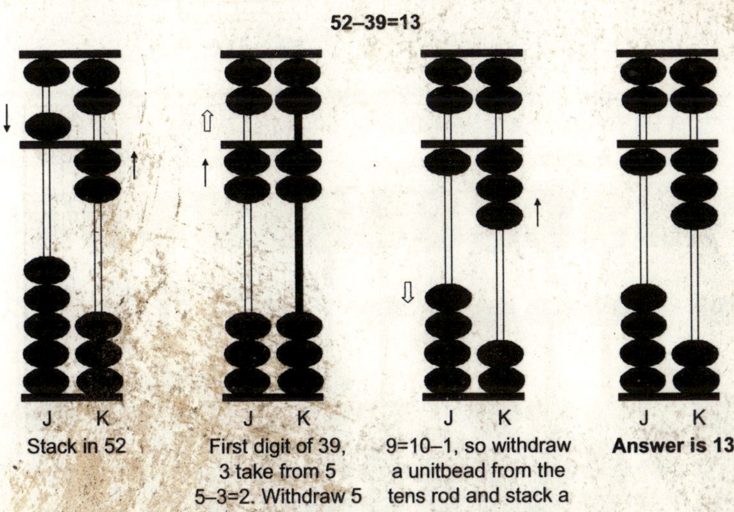

Taking-off from a Rod of Higher Order and Adding-up

When a number on a specific rod is smaller than the subtrahend (4 is the subtrahend when performing 13–4). Note that in the ones column, the 3 is less than the 4. So, take off one bead for the order of tens and move one bead up in the lower deck and move down one bead in the upper deck from the rod which means an actual reduction of 4, and gives the correct result as 9.

Given the first number	*To add (formula)*	*Move bead(s)*		
		In the lower deck	*In the upper deck*	*Lower deck, adjacent (left) column*
0	1(+9–10)	+4	+1	–1
0 or 91	2(+8–10)	+3	+1	–1
0, 1 or 2	3(+7–10)	+2	+1	–1
0, 1, 2 or 9	4(+6–10)	+1	+1	–1
0, 1, 2, 3 or 4	5(+5–10)		+1	–1
0 or 5	6(+4–10)	+4		–1
0, 5 or 6	7(+3–10)	+3		–1
0, 1, 2, 5, 6 or 7	8(+2–10)	+2		–1
0, 1, 2, 3, 5, 6, 7 or 8	9(+1–10)	+1		–1

Example 8: Find 10,111,210,200,20–1,020,304,567,89.

This subtraction is explained in following Figures.

Fig. 1

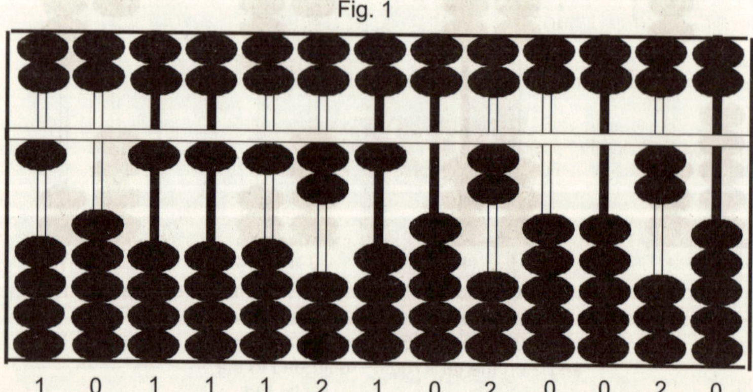

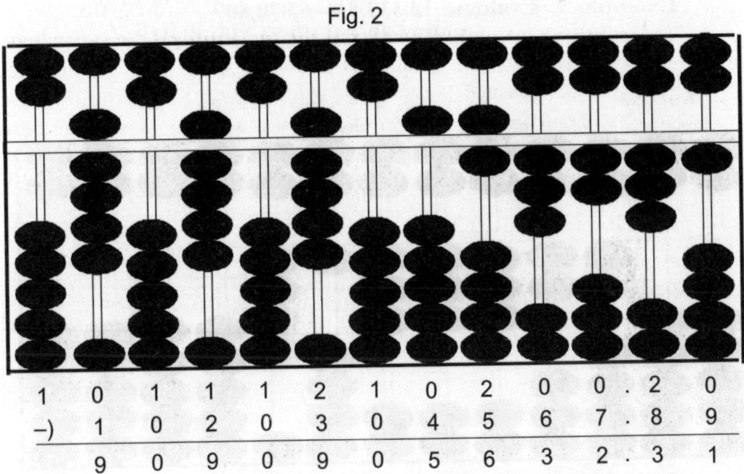

Fig. 2

Combined taking-off from a rod of Higher order, Adding-up in the Upper deck and taking-off in the Lower deck—This technique is called for when the number on a certain rod is smaller than supposed subtrahend, but only in each cases as may be exemplified by 12–6. To carry out this subtraction, intakes off one bead from the rod for the order of tens, moves down one bead as unit place in the upper deck, and again takes off one bead counter in the lower deck, leaving the number 6 on the unit rod, which is the required answer. That this is correct can be confirmed by the fact that the algebraic sum resulted from the operation performed as above is 6, the subtracter.

Given the first number	*To add (formula)*	*Move bead(s)*		
		Lower deck, adjacent (right) column	*In the lower deck*	*In the upper deck*
2, 2, 3 or 4	6(–1+5–10)	–1	–1	+1
2, 3 or 4	7(–2+5–10)	–1	–2	+1
4 or 4	8(–3+5–10)	–1	–3	+1
4	9(–4+5–10)	–1	–4	+1

Example 9: Evaluate 12,333,314–6,070,809.
The subtraction method is shown below in Figures 1 and 2.

Fig. 1

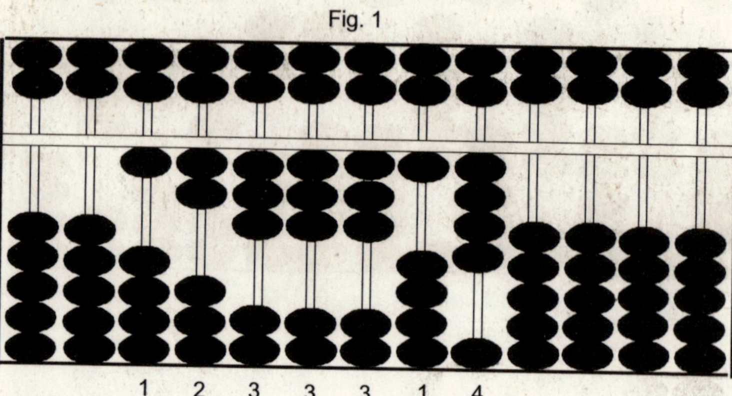

Fig. 2

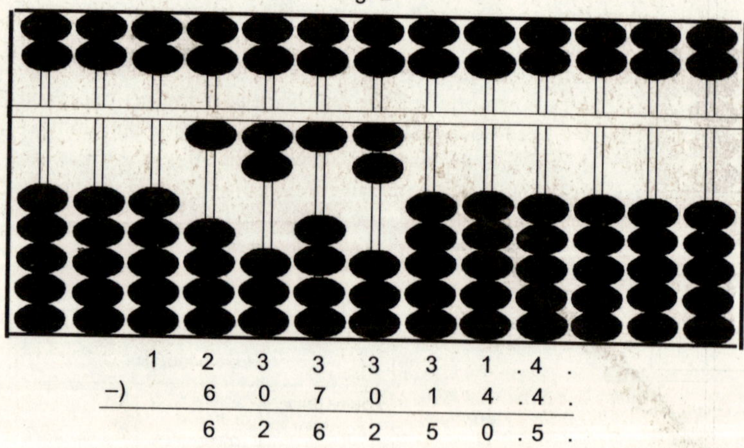

To carry out a series of subtractions. It is obviously easier to use an abacus than to make the calculations on paper.

The actual procedures can be shown as follows:

Original minuend

–) 1st subtractor
1st remainder

–) 2nd subtractor
2nd remainder

–) 3rd subtractor
3rd remainder

–) 4th subtractor
4th remainder

===== :
===== :
===== :

(n–1)th remainder

–) nth subtractor
nth remainder (Final result)

Consider an Example: 3,573,509–1,321,402–443,086–693,715 =1,115,306. This is described in the following four steps (Figures).

Fig. 1

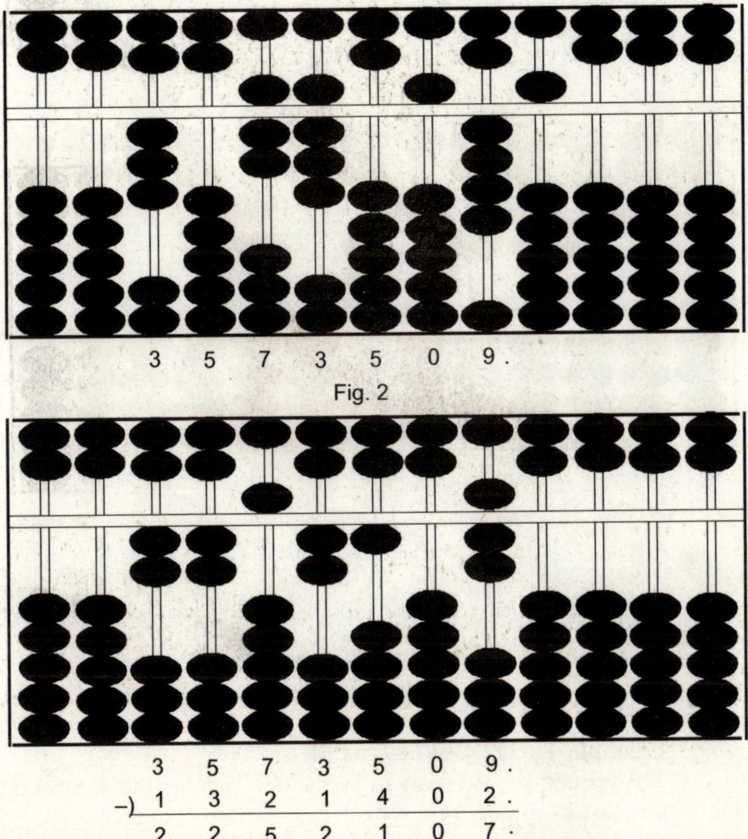

Fig. 3

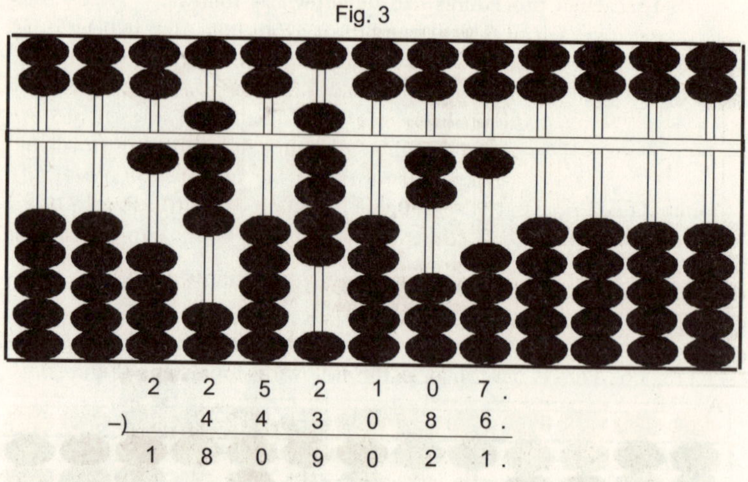

```
    2  2  5  2  1  0  7.
-)     4  4  3  0  8  6.
   ---------------------
    1  8  0  9  0  2  1.
```

Fig. 4

```
    1  8  0  9  0  2  1.
-)     6  9  3  7  1  5.
   ---------------------
    1  1  1  5  3  0  6.
```

Notice that in carrying out additions and subtractions, one does not have to use the two auxiliary operation fields at all.

DECIMALS

Example 10: 102.43–67.89=34.54

The vernier: choose rod I to be the unit rod. Set the decimal on the vernier just to the right of rod I.

Step 1: Set 102.43 onto rods GHIJK.

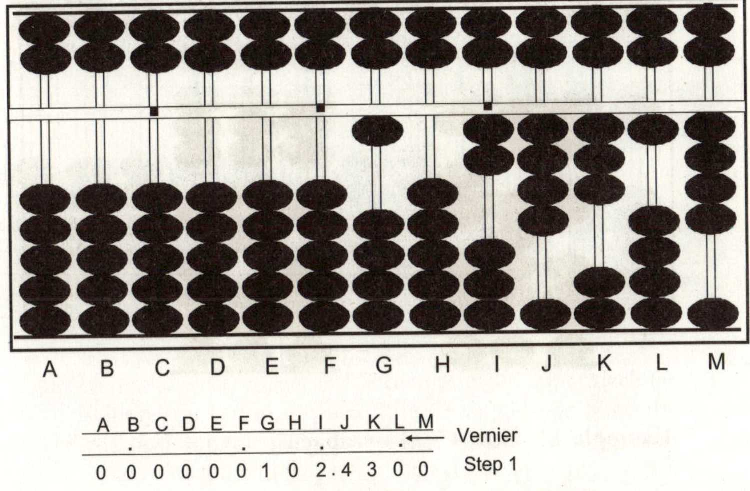

Step 2: Solve the subtraction using complementary numbers where necessary just as you would on a normal suanpan. This leaves the answer 34.54 on rods HIJK

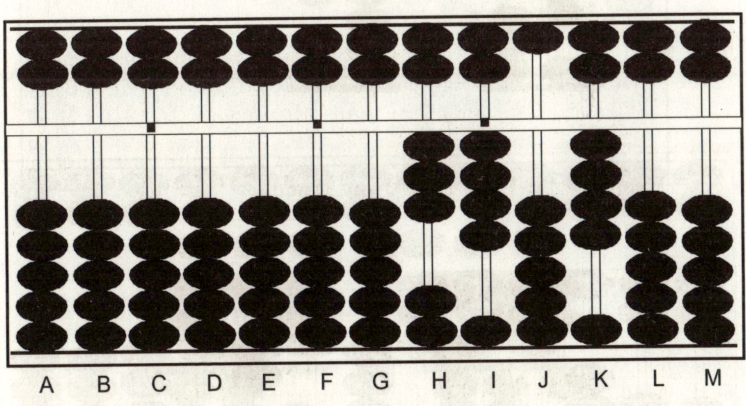

A B C D E F G H I J K L M

0 0 0 0 0 0 1 0 2.4 3 0 0
− 6 7.8 9 Step 2

0 0 0 0 0 0 0 3 4.5 4 0 0

MORE EXAMPLES

Example 11: Find 146–25 on abacus

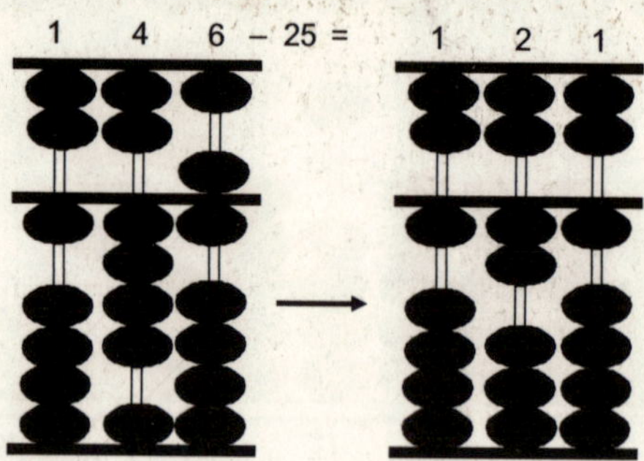

Example 12: Find 47–34 on abacus

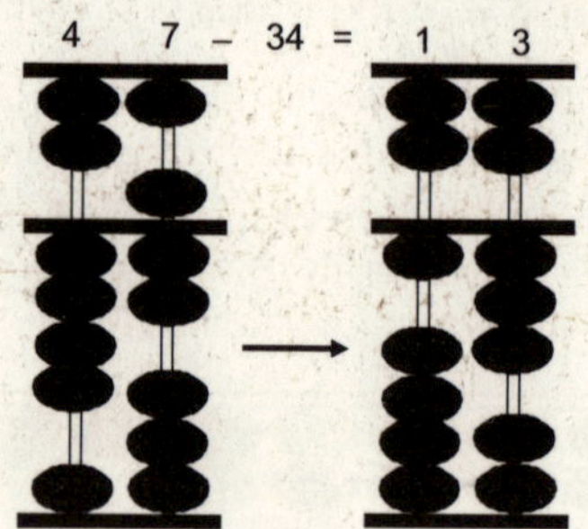

Example 13: Find 128–36 on abacus

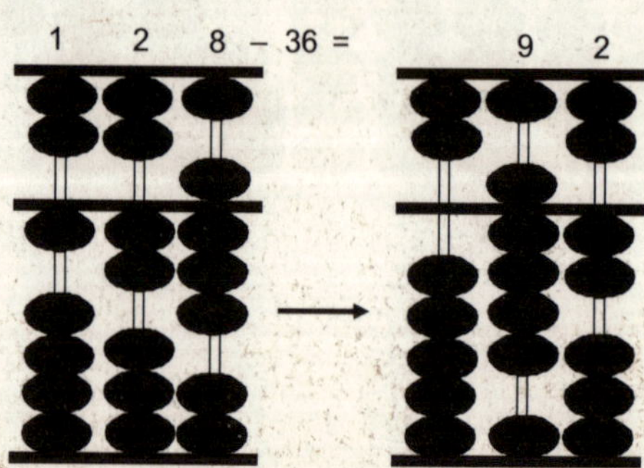

Example 14: Find 623–24 on abacus

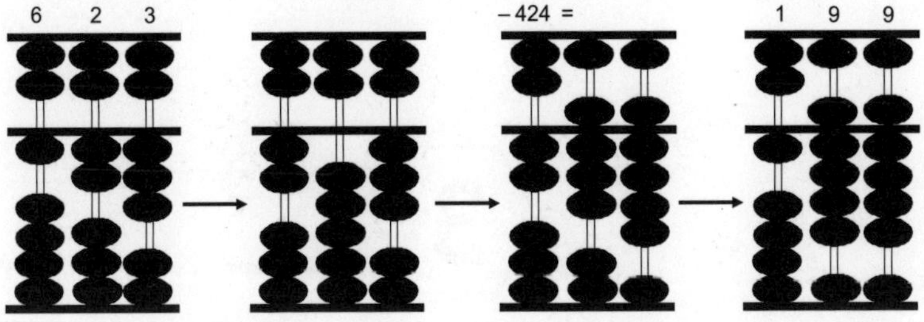

Note: When subtracting, we are riot simply removing beads hot sometimes adding beads, in mathematical expression, these operations are as follows:

–2=–5+3
–2=–10+8
–6=–10+4
–6=–1+5–10

Experienced users can use two or three fingers (thumb, index, or middle finger) to move the beads. For example, 8+4=6+10, 6 can be done in two moves. The first move adds 1 to the rod on the left using thumb or index finger; the second move subtracts 6 using both thumb and index finger (or middle finger) by pushing beads outward. Similarly +3=+5, 2 can be done in one move: pushing down one heaven bead with index (Or middle) finger and 2 earth beads with thumb simultaneously. Alternatively +3+5, 2 can be done in one continual move: using the index finger to push down one heaven bead and continue on to push down 2 earth beads.

Multiplying on an Abacus

To start, you need to know how to write numbers on the Chinese abacus. Numbers are written with the is column on the far right, then the 10s column, then the 100s column, etc. and you only look at which beads are touching the center bar. Beads on the top count as 5, while beads on the bottom count as 1. Here's the number 358:

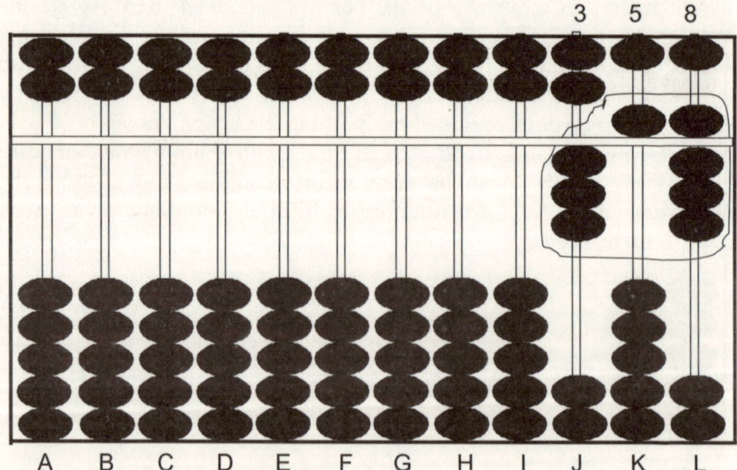

Notice that you could get by with each rod only having 1 bead on the top and 4 on the bottom. On to adding! Some of this is easy: If you have a number like 358 and you want to add 100, you just move one of the 100 beads [third column from the right].

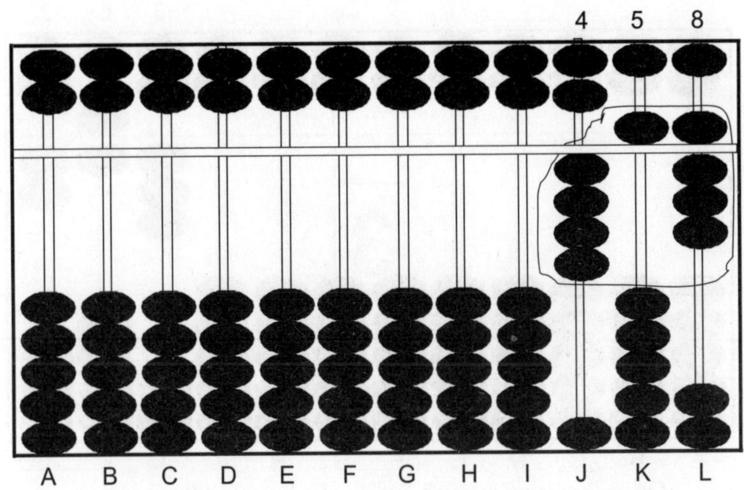

But what if it's not that simple? If you have a number like 358 and you want to add 18, you could do it bead by bead *(Slow.* Boring. Lots of regrouping.) or you could just add 2 tens and then remove 2 ones. That's the more authentic way.

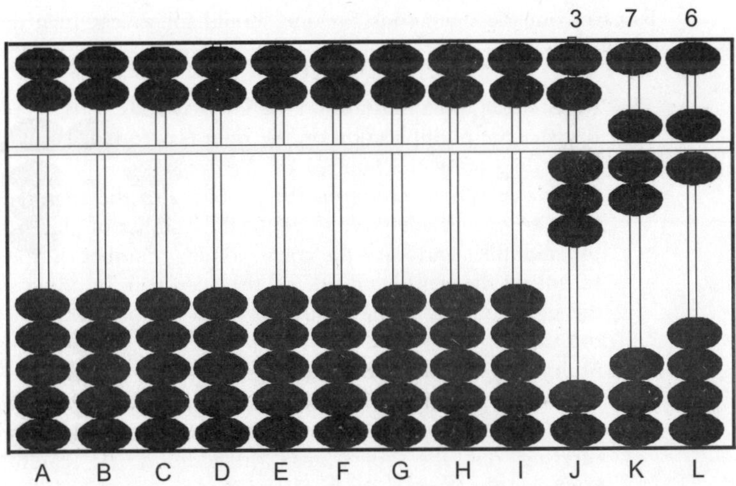

Likewise, if you had 358 and you wanted to add 4, you could add a five bead and subtract 1 unit, then ,regroup, or you could add 1 ten bead and subtract the five and a one bead.

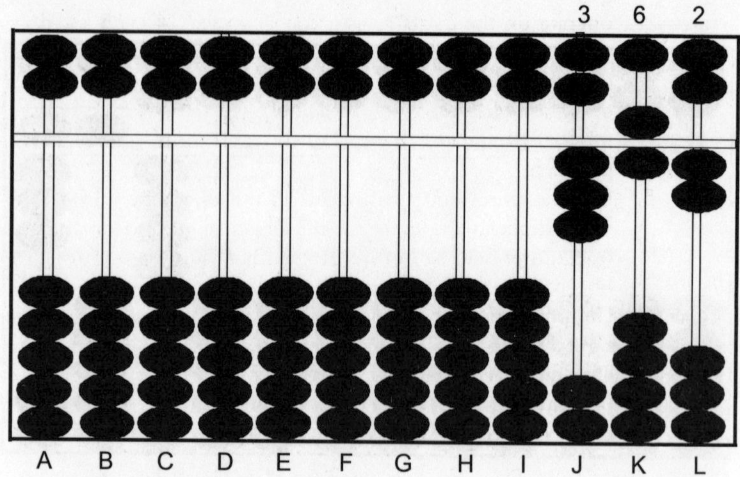

On to multiplication? For single digits, well, you pretty much just need to know all your basic multiplication facts. No way around it. But this process will work for anything beyond that.

The method of multiplication on the abacus is similar to the method used to ordinary arithmetic except for the fact that it can be carried out more rapidly, once the user get used to it.

To carry out the operations for one should follow carefully the steps prescribed below:

1. Register the multiplication on the left (say rods A, B, C, D)
2. Register the multiplication on the right (say rods I, J, K, ...) depending upon the number of digits.
   Please note: When setting up the problems in the examples given below, it's advisable to count the number of digits in the multiplier and leave the corresponding number of rods vacant on the right hand side of the Suan Pan. In doing so the unit number in the product will fall neatly on the last rod.
3. Multiplying the digits on the left rod by rod, in a right-to-left order by the right-end digit on the right and register the products one after another on the operations field as in carrying out the operations for addition. Be careful, however, not to place the bead counters on wrong rods, and always, remember to move one-rod to the left in registering the next product. The indicate should be consulted as a reliable guide for placement.
4. Use the second digit from right on the right as multiplier to

operate on the digits registered on the left, rod by rod, in a right-to-left order, and register the products on the operations field one after another as in carrying out operations for addition. To avoid misplacing bead counters, move the indicator one rod to the left before adding up the products.

5. Use the third digit from right on the right operations field as multiplier to operate on the digits registered on the left operations field as before. Remember to move the indicate one rod to the left to avoid misplacing.
6. Continue the operations as above until every digit on the right operations field has been used once.
7. The final reading on the operations field gives the required product and the place setting vernier now indicates the correct places for all digits.

Before we start, let us explain what multiplier and multiplicand are. In the expression

a*b

a is the multiplier and *b* is the multiplicand.

The steps of multiplication on abacus are summarized in the figure below.

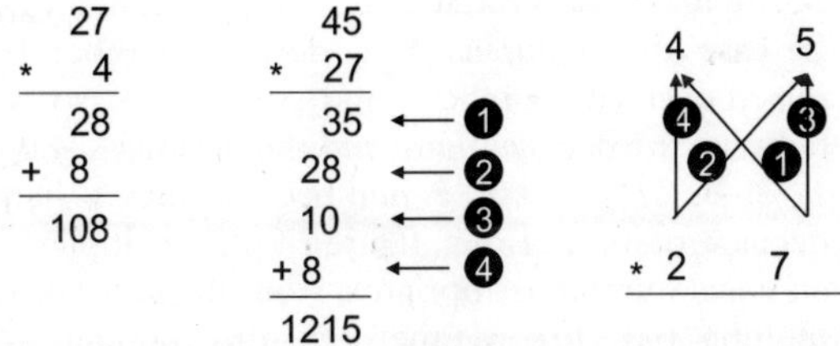

Step 1. Start from right to left for each digit of the multiplicand, calculate the multiplication product for the digit and a digit in the multiplier. The order for the multiplier is also from right to left.

Step 2. Add the multiplication product number on the abacus.

Step 3. As one moves from right to left for either multiplicand or multiplier, the product number should be added on the abacus one column to the left.

It is confusing to read these sentences. Let us go through some examples. The procedure is actually very simple.

Example 1: To prove 1.27 * 4=108 on abacus.

Step 1. The first digit from right for the multiplicand is 4. The first digit from right for the multiplier (27) is 7. So, 4 * 7 = 28. We add 28 on the abacus.

Step 2. The first digit from right for the multiplicand is 4. The second digit from right for the multiplier (27) is 2. 4 * 2 8. Since we have moved one digit to the left for the multiplier, so the number 8 should be added one column left relative to the unit column. That is we need to add 80 on the abacus. The result is 108.

Determination of unit position in the product:

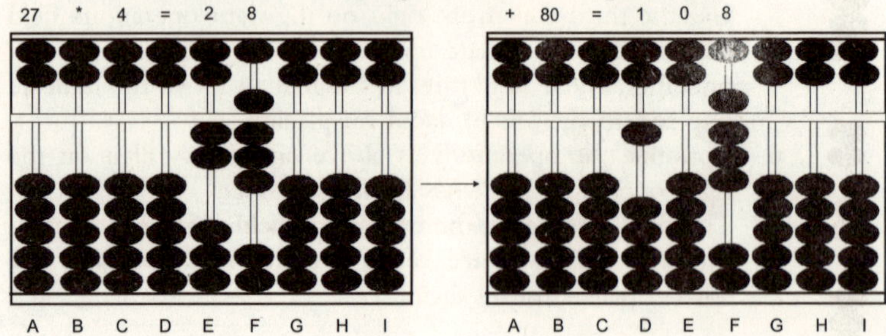

(1) In such case where the multiplier is the integral or mixed decimal number (for example 3.56; 19.362 etc), count figures in the integral part of the multiplier. Count off equal number of uprights, from the unit Position of the multiplicand, toward the right: this upright is the unit position for the product.

(2) In case the multiplier is a decimal number (except the mixed-decimal number) and there are some "Zeros" between the decimal point and the decimal-significant figure 1, 2, 3, ...9. Count a number of "zeros" between the decimal point and the figure of the multiplier. Count off an equal number of uprights, from the unit Position of the multiplicand, toward the left: this upright is the unit Position for the product.

(3) In such case where the multiplier is the decimal number and there is no "zero" between the decimal point and the number the unit Position of the multiplicand is the unit

Example 2: Show that 67×2=134 on abacus.

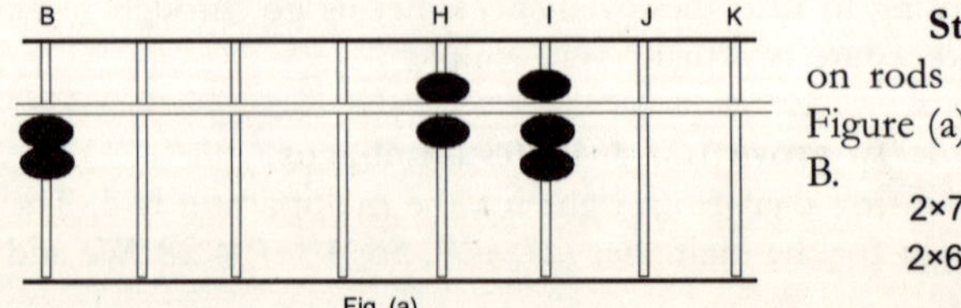

Fig. (a)

Step 1: First 67 on rods HI as shown in Figure (a) and two on rod B.

2×7= 14
2×6=12
134

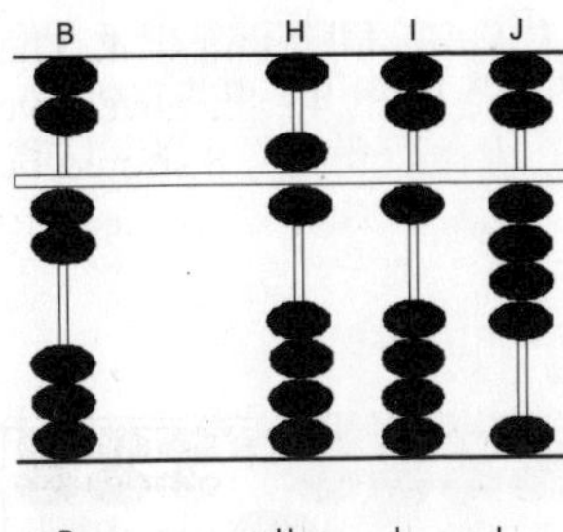

Step 2: Compare the last figure of the multiplicand, i.e., 7 on the upright I with the multiplier 2. As, 2×7 is 14, therefore, after taking off 7 from the upright I place 1, the first figure of 14, on the upright I and place 4, the second figure, on the upright J (Fig. (b)). Next, compare 6, another figure in the multiplicand, with the multiplier, 2.

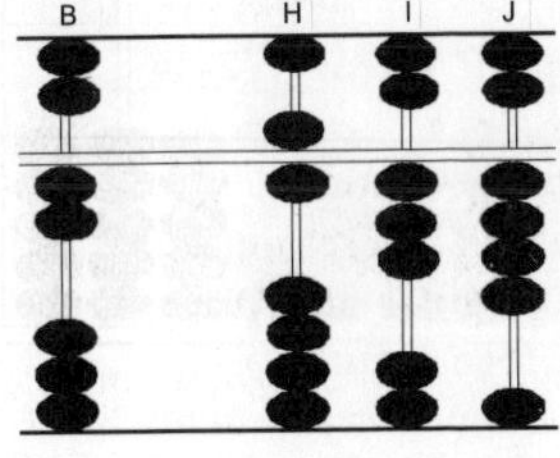

Step 3: "2×6 is 12" therefore, after taking off 6 from the upright H place 1, the first figure of 12, on the upright H, then add 2, the second figure, to 1 on the upright I, The product is 134 (Fig. (c)).

Example: 9×7=63

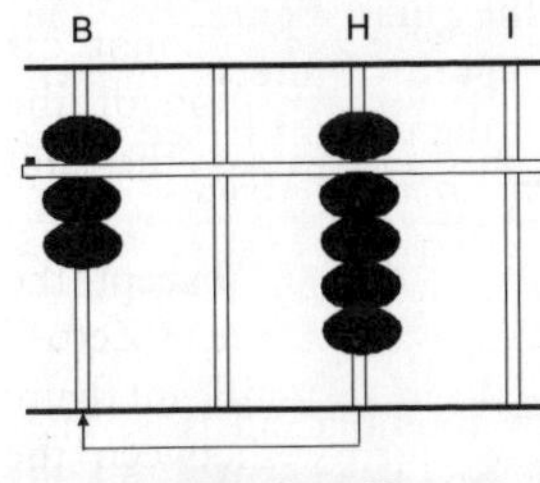

Compare the multiplicand, i.e. 7 with the multipliers is 9 "9×7 is 63", therefore, take off 9 and place 6, the first figure of 63, on the upright H and place 3, the second figure on the upright. I after taking off 9 from the upright H. The product is 63.

Answer 63

Example 3: Show 14×26=364 on abacus.

We proceed similarly as we did in the above example.

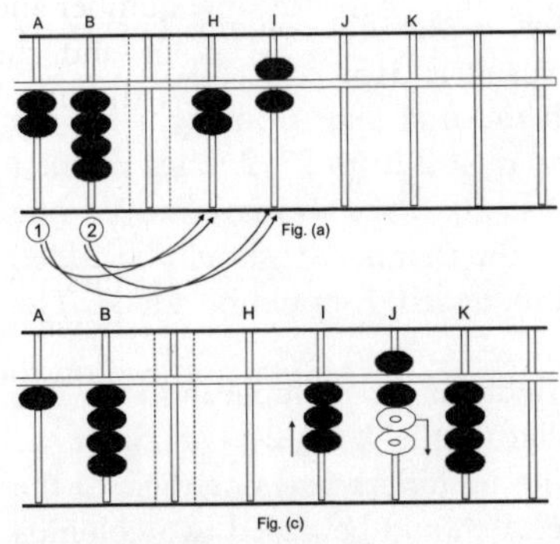

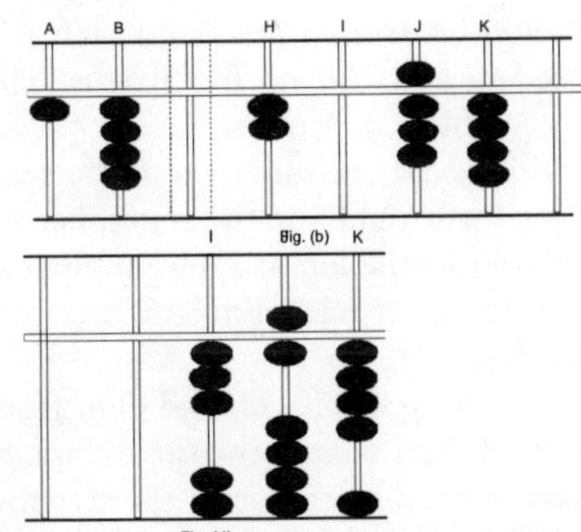

In illustration above Fig. (a), we see that the multiplier is a two place integral. Therefore, count off two uprights, from the unit position of the multiplicand, i.e., from the uprights I toward the right. The upright K is the unit position for product.

Process of Multiplication

(1) Compare the last figure of the multiplicand, i.e., 6 on the upright I with the first figure of the multiplier, i.e. 1.
"1×6 is 6", therefore, after taking off 6 from the upright I place 6 on the upright J.

6×1=	06
6×4=	24
2×1=	02
2×4=	08
	364

(2) Next, compare the same 6 with the second figure of the multiplier. 4, "4×6 is 24", therefore, add 2, the first figure of 24 to 6 on the upright J and place 4, the second figure, on the upright K (See Figure (b)).

(3) Compare 2, the first figure of the multiplicand, with 1, the first figure of the multiplier. "1×2 is 2", therefore, place 2 on the upright I, after taking off 2 from the upright H. Finally, compare the same 2, the first figure of the multiplicand, with 2, the second figure of the multiplier, "4×2 is 8", therefore, add 8 to 8 on the upright J. (see (Fig. (c)). The product is 364 as shown on the "Abacus" now. (Fig. (d)).

Example 4: Prove 45×27=1215

Step 1: The first digit from right for the multiplicand is 7. The first digit from right for the multiplier (45) is 5. So, 7×5=5. We add 35 on the abacus.

Step 2: The first digit from right for the multiplicand is 7. The second digit from right for the multiplier (45) is 4. So 7×4=28. Since we have moved one digit to the right for the multiplier, so the number 28 should be added one column left relative to the unit column. That is we need to add 280 on the abacus. The result is 315.

Step 3: Now we are done with the first digit from right for the multiplicand, we move on to the second digit 2 from 27. 2 times the first digit from right for the multiplier (45) is 10. Since we have moved one column to the left for the multiplicand, the number 10 should be added 1 column left relative-to the unit column, i.e., 100 should be added. The result is 415.

Step 4: The second digit from right for the multiplicand is 2. The second digit from right for the multiplier (45) is 4. So, 2×4=8. Since we have moved one digit to the left for the multiplier and one digit to the left for the multiplicand, so the number 8 should be added two columns left relative to the unit column. That is we need to add 800 on the abacus. The final result is 1215.

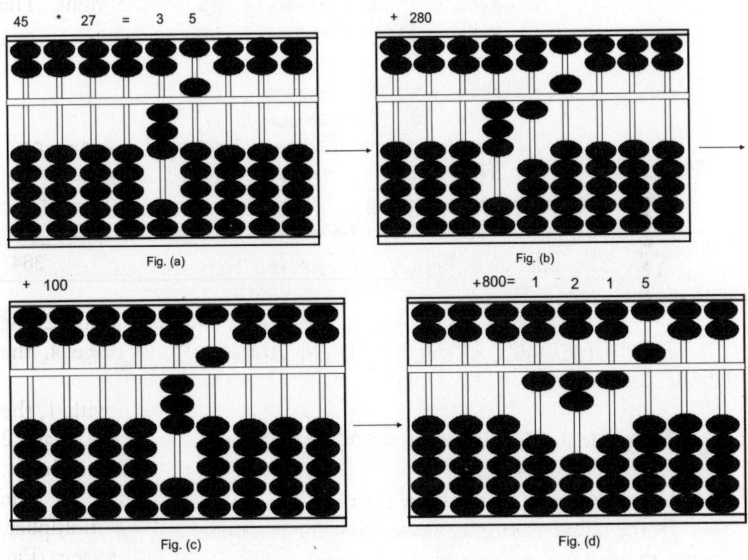

Fig. (a) Fig. (b) Fig. (c) Fig. (d)

Hence, **45×27=1215**

Example 5: Show that 345×678=233910

Determination of the unit position for the product: in this case, the multiplier is a three place integral, therefore, count off three uprights, from the unit position of the multiplicand, i.e., from the upright J toward the right; the upright M is in the unit position for the product. The diagram above shows the different steps on "Abacus" in the course of calculation of "678×345".

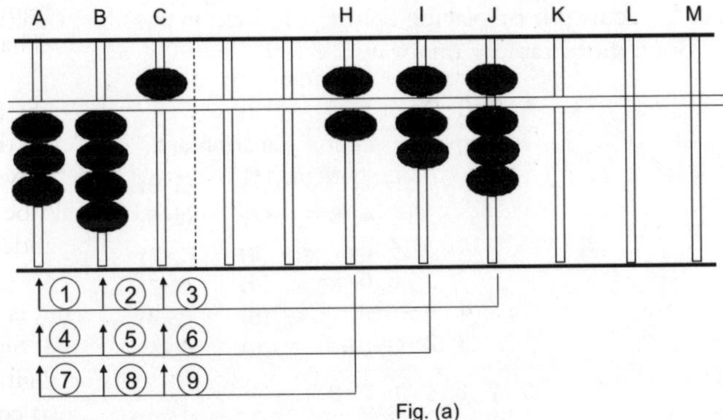

Fig. (a)

8×3=	24	======	(1)
8×4=	32	======	(2)
8×5=	40	======	(3)
7×3=	21	======	(4)
7×4=	28	======	(5)
7×5=	35	======	(6)
6×3=	18	======	(7)
6×4=	24	======	(8)
6×5=	30	======	(9)
Required answer==	233910		

Fig. (a)

Note: The encircled figures indicate the steps in calculation in Figure (a).

DETAILED WORKING IS EXPLAINED IN EXAMPLES BELOW:

Example 6: To calculate 4.076×0.028

Here, the multiplier is 0.028

Determination of the unit position for the product.

In the above example, we see that the multiplier has one 0 "zero" between the decimal point and the decimal-significant figure. Therefore, count off one upright from the unit position of the multiplicand, i.e., from the upright B, towards the left: Then the upright A is the unit position for the product.

□ **4.076×0.028=0.11428**

Note: The encircled figures indicate the steps in calculation.

Note: At first, calculate by placing both "multiplicand" and "multiplier" on the "Abacus". As you practice on the "Abacus", calculate by placing either the "multiplicand" or multiplier" on the "Abacus". If you practice further and become proficient in calculation on the "Abacus", calculation by placing neither of them is possible. Calculation can be done more rapidly this way.

A	B	C	D	E	F	G			
	4	0	7	6			=== ..	multiplicand	
				1	2		== ..	(1)	(2×6)
					4	8	=..	(2)	(8×6)
			1	4			== ..=..	(3)	(2×7)
				5	6		== ...	(4)	(8×7)
		8					====== ..	(5)	(2×4)
		3	2				===== ..	(6)	(8×4)
0	1	1	4	1	2	8			

Note: The following steps before multiplying on Chinese abacus, example is to multiply 713×289=?

713×289=?

The integer multiplicand 713 is multiplied by the multiplier 289. We must first set out space on the abacus for our calculation.

- If a single digit number is multiplied by another single digit nudiber, the maximum number of digits the product has is two.
- If a single digit number is multiplied by a double digit number, the maximum digits the product has is three.
- If both the multiplicand and multiplier are double digit, their product has a maximum of four digits.
- The trend is that for an M digit multiplicand which is multiplied by an N digit multiplier, the product has a maximum of (M+N) digit result.

In our example, both 713 and 289 have three digits each so that makes a maximum number of 6 digits in the resultant product. Hence, for the calculation, six rods on the abacus-is needed.

In this case, count three rods along from the right side of the abacus, then another three and place the multiplier in the latter three rods. We shall see that the production will eventually replace the multiplier as each of its digits are considered in the calculation. Some where away from the six rods, you place your multiplicand for reference.

The process of multiplication requires the learner to start with the rightmost digit of the multiplier, and multiply the multiplicand digit by digit from the left of the multiplicand to the right. The second rightmost digit of the multiplier then repeats the process with thee muftiplicand in the same manner until the leftmost digit of the multiplier is done. The product will be that which is shown at the end of the manipulation.

So, the calculation follows the following route.

Example 7: To find 713×289

Step 1: The right most digit of the multiplier is "9". Multiply the leftmost digit of the multiplicand, "7".

Step 2: Next, 9 multiplies the next digit of the multiplicand, "1".

Step 3: Then. 9 multiplies the last-digit of the multiplicand, "3".

We then repeat this process with the second digit of the multiplier "8", and finally the leftmost digit "2".

As we said, the manipulation of the beads on the abacus requires one to think about two rods at a time. For example, in step (1), the product of this minor calculation was 63. We note the position of digit nine of the multiplier and the rod to its right. This forms the tens and units place in which the numeral 63 is inserted replacing the digit 9 by the 6 and placing 3 to the rod on its right. In step 2, the product is nine, a single digit. Again, we consider this as a double digit number 09, and insert increment 0 beads in the tens (where the digit 3 appears), and put in the value of 9 on the abacus to the next rod on the right. In step 3, the product of 9 by 3 is 27. This move requires us to consider more than two rods. This is because we must put in the value 2 to the position of 9 on the rods, but this would make II, hence the a bead on the next left rod is incremented by one, and the rod containing 9 would be amended to 1 and finally the digit 7 would be put into the rightmost column as the units column.

The next step with the second digit of the multiplier, viz, "8", would give the products 56. 8 and 24. The first digit of 56, "5" replaces the multiplier digit "8" and the second digit is added to the value on the rod immediately to the right, and so on.

Finally, the products of the last digit of the multiplier, viz, "2", yields the products, 14, 2, 6. The first digit of 14, "1", replaces the first digit of the original multiplier, and the second digit is added to the rod immediately to its right. The other numbers are placed in in the same way described above.

What we have is the following representation

*** are unused rods.

--- are places which have not been considered yet,
but are about to change because of carrying overflow.

+++ are digits which are not part of the final product

```
713  289***  Abacus is loaded thus     Abacus reads 289***
                                       ++
      63     The product of 9 and 7    Abacus reads 2863**
        9    The product of 9 and 1    Abacus reads 28639*
       27    The product of 9 and 3    Abacus reads 286417
             +
     56      The product of 8 and 7    Abacus reads 25...then 262417
      8      The product of 8 and 1    Abacus reads 262.then 263217
      24     The product of 8 and 3    Abacus reads 263457
   14        The product of 2 and 7    Abacus reads 1 then 203457
      2      The product of 2 and 1    Abacus reads 205457
       6     The product of 2 and 3    Abacus reads 205.then 206057
```

The answer is the product **206057**.
You can check every step on the abacus.

Continuing with different ways to multiply, In the examples below there are three numbers in the multiplier, so the last three rods on the right are left vacant.

Example 8: Find 87×625

To find 87×625. First pick one of the numbers (87) and put it on the far left-hand side of the abacus in columns A and B (see Figure below). Since 87 is a 2 digit number, we want to leave 3 blank spaces that's (2+1) on the far right in columns K, L, M and we'll put 625 after that (where by "after" I guess I mean "before", to the left) in columns H, I, J, Just look at the picture and it should be clear.

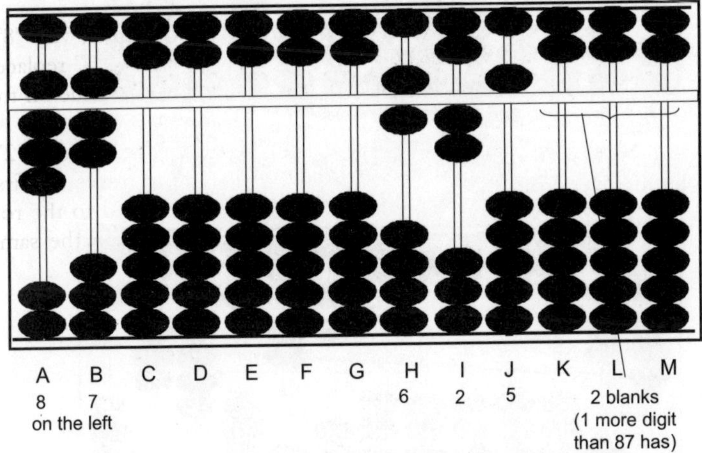

Before we get started on the process, it's helpful to take a look at the big picture. We're going to put our answer on the far right, which is why we needed those blank spaces. But our answer will be more than 3 digits long, so we're going to need more space. We'll create the extra space as we move along. We'll first multiply 87 by the 5 of 625 then we'll get rid of that 5, freeing up an extra column. then we'll multiply 87 by the 2 of 625, freeing up yet another column, and finally we'll multiply the 87 by the 6.

Let's get started.

We want to multiply 87 by the 5. We'll multiply the digit 8 and then the digit 7 [left to right]. Yes, the is backwards from how we're multiplying 625 [right to left], but that just keeps it fun. Whenever we multiply by the of 7 we'll put the answer two columns over, and whenever we multiply by the 7 of 87 we'll put the answer three columns over. [No matter how many digits that number is, when you multiply by the largest place-value you put the answer two columns over; the next place value goes three columns over; the next would go four columns over, etc.]

So to start, we know that 8x5 is 40. Since we're using the 8, we'll put the answer two columns over from the 5. We always use two for the left-most column.

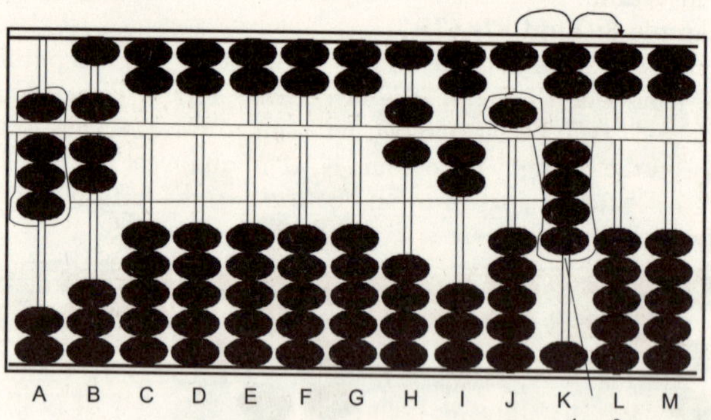

Now we'll move to 7x5 This is 35, and we put the answer three columns over from the 5.

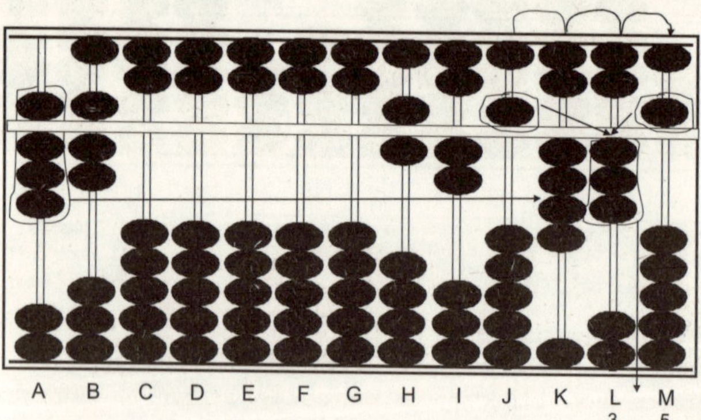

Since we're done multiplying 87 by 5, we can remove the 5.

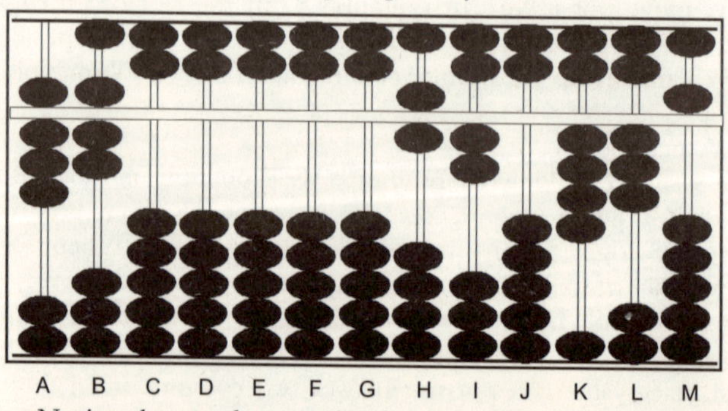

Notice the number on the far right, 435, is just 87×5.

We'll do the same thing, multiplying 87 by 2. First we find 8x2 and put that answer two columns over from the 2. [I'm going to do that by adding 20 instead of 16. and then subttation 4].

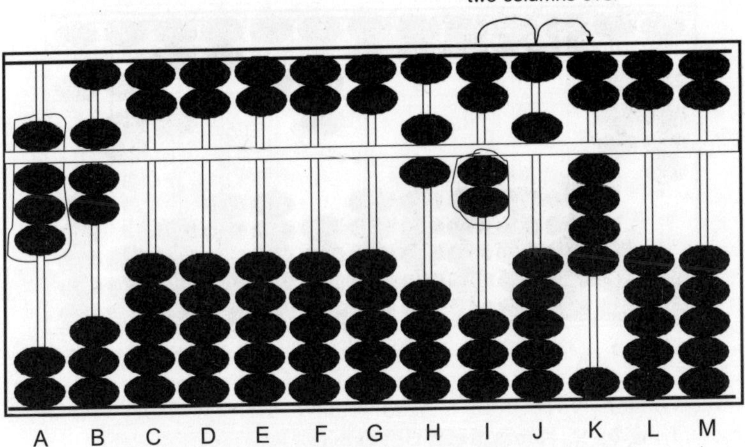

Then we'll multiply 7×2 and put that answer three columns over from the 2. [Here I added 15 instead of 14, and then I subtracted 1.]

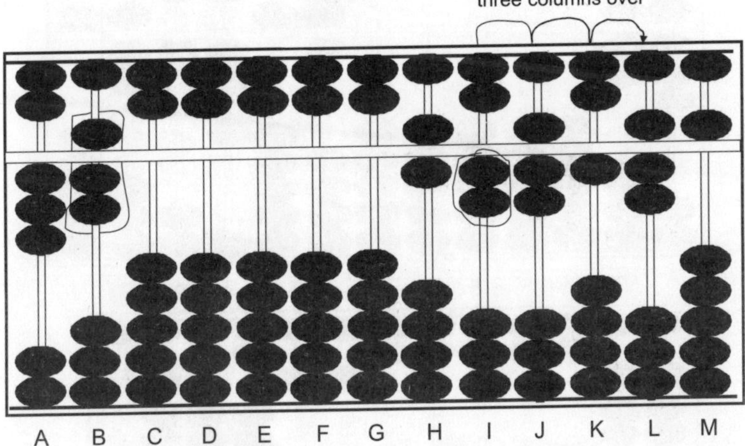

We're done multiplying by 2, so we can remove tt. Lxcept I forgot to take a picture with it removed. If you ignore that 2 of 62 (formerly 625), you can see that the number on the far right is now 2175. Sure enough, this is 87×25.

Nearly done, we'll multiply the 8 of 87 by 6 and put the answer two columns over from 6. Notice that this 6 originally stood for 600 (in

625), but we don't need to keep track of which place value the 6 originally had. By always putting the single-digit products two or three (or four for longer numbers) over, the place value takes care of itself.

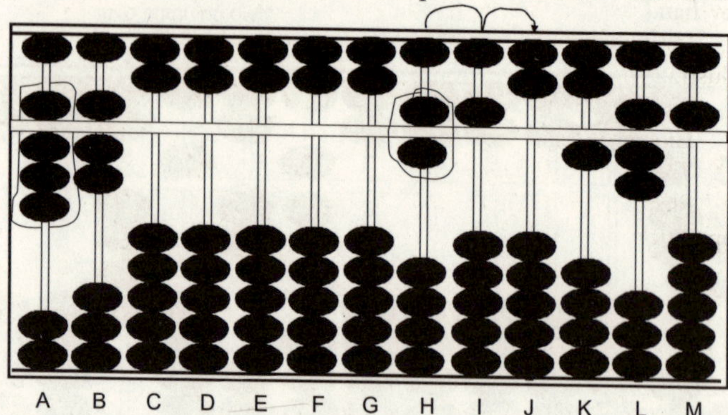

And finally, we multiply the 7 of 87 by the 6 and put the answer three columns over from 6. And remove the 6 because we're done multiplying by it. Here's the final answer:

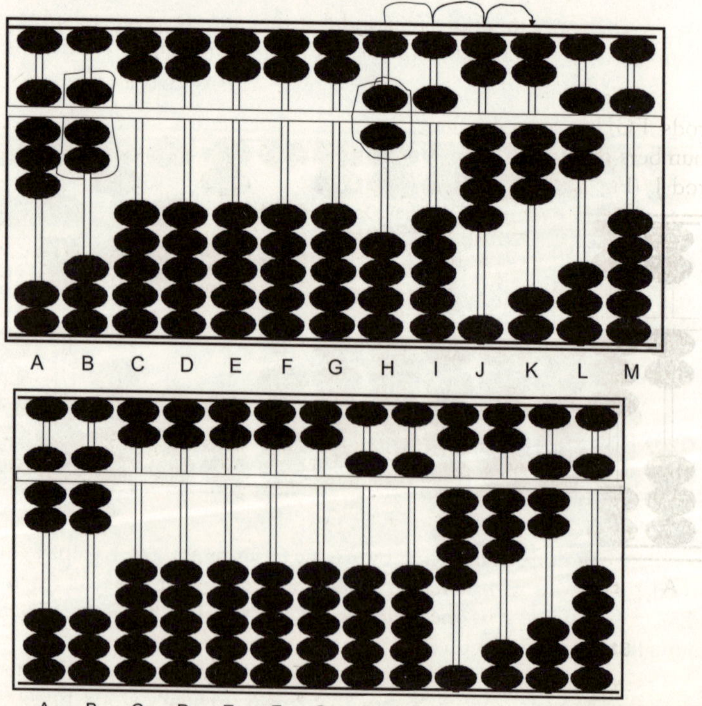

Yes, it's 54,375. And a quick check shows that It's right!

And despite how long it takes to explain, it really isn't too bad to do by hand even if you're learning it for the first time.

Let us consider the following two simple examples that will help to illustrate how this traditional multiplication method works.

Example 9: Prove that 456×23=10488 on abacus.

The Order of Multiplication:

Please refer to Fig. 1 below to help with the illustration. The order of operation is as follows:

- Rod H×B×A.
  Note: the multiplicand on rod H changes to become part of the product.
- Rod G×B×A.
  Note: the multiplicand on rod F changes to become part of the product.
- Rod F×B×A.
  Note: the multiplicand on rod G changes to become part of the product.

Step 1: Choose rod H to be the unit rod. Set multiplicand 456 on rods FGH and multiplier 23 on AB. The multiplier has two whole numbers therefore the unit rod will shift two to the right bringing us to rod J. (Fig. 1)

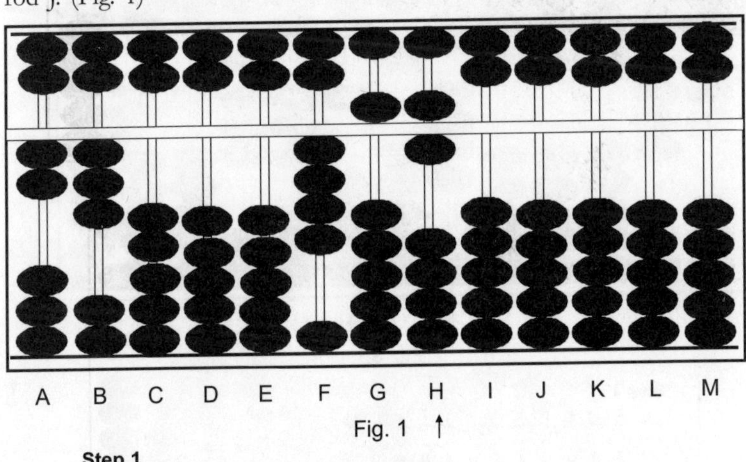

Fig. 1

Step 1

A	B	C	D	E	F	G	H	I	J	K	L	M
2	3	0	0	0	4	5	6	0	0	0	0	0

Step 2: Multiply 6 by 3 and add 18 to rods IJ. (Fig. 2)

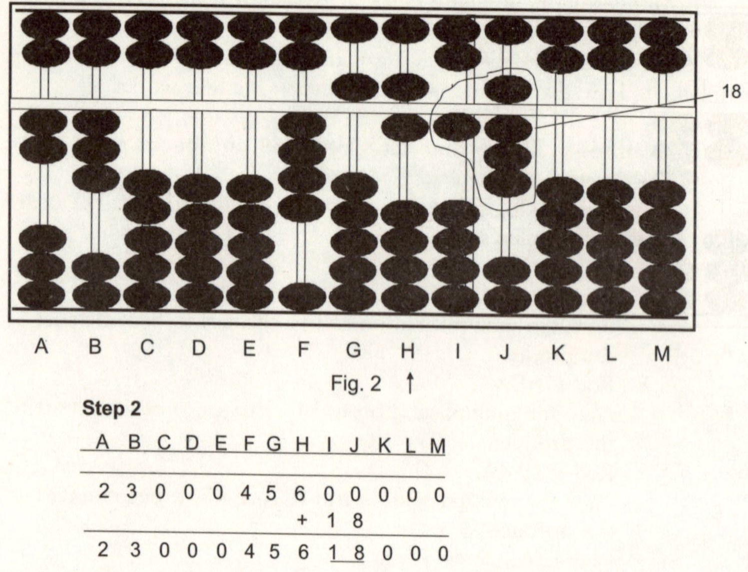

Fig. 2

Step 2

A	B	C	D	E	F	G	H	I	J	K	L	M
2	3	0	0	0	4	5	6	0	0	0	0	0
							+	1	8			
2	3	0	0	0	4	5	6	<u>1</u>	<u>8</u>	0	0	0

Step 3: Multiply 6 by 2 and add 12 on rods HI. Note that what was multiplicand 6 on H has now changed to 1. (Fig. 3)

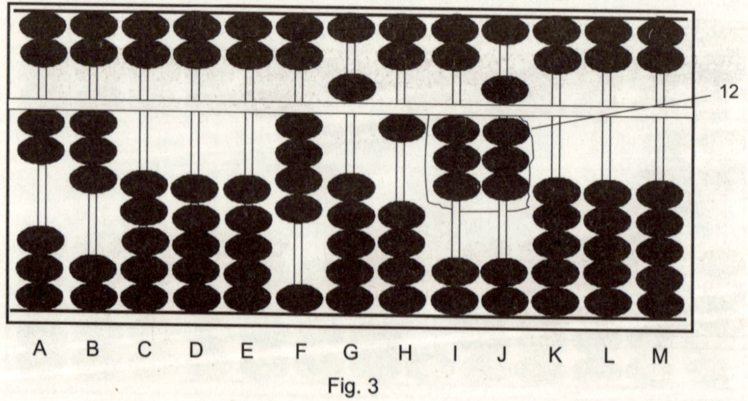

Fig. 3

Step 3

A	B	C	D	E	F	G	H	I	J	K	L	M	
2	3	0	0	0	4	5	6	1	8	0	0	0	
						+	(1)	2					←
2	3	0	0	0	4	5	<u>1</u>	<u>3</u>	<u>8</u>	0	0	0	←

Note: Difference in steps marked with ←

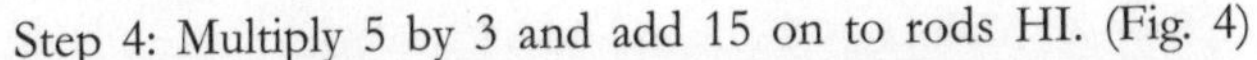

Step 4: Multiply 5 by 3 and add 15 on to rods HI. (Fig. 4)

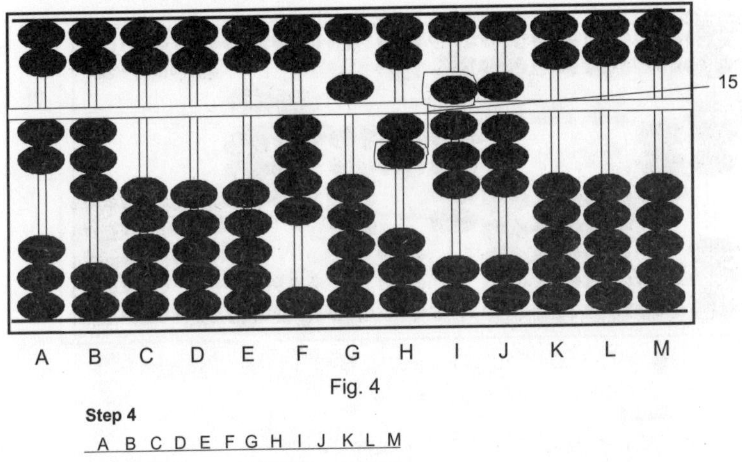

Fig. 4

Step 4

A	B	C	D	E	F	G	H	I	J	K	L	M	
2	3	0	0	0	4	5	1	3	8	0	0	0	
						+	1	5					←
2	3	0	0	0	4	5	2	8	8	0	0	0	←

Note: Difference in steps marked with ←

Step 5: Multiply 5 by 2 and add 10 to rods GH. Note that what was multiplicand 5 on G has now changed to 1. (Fig. 5)

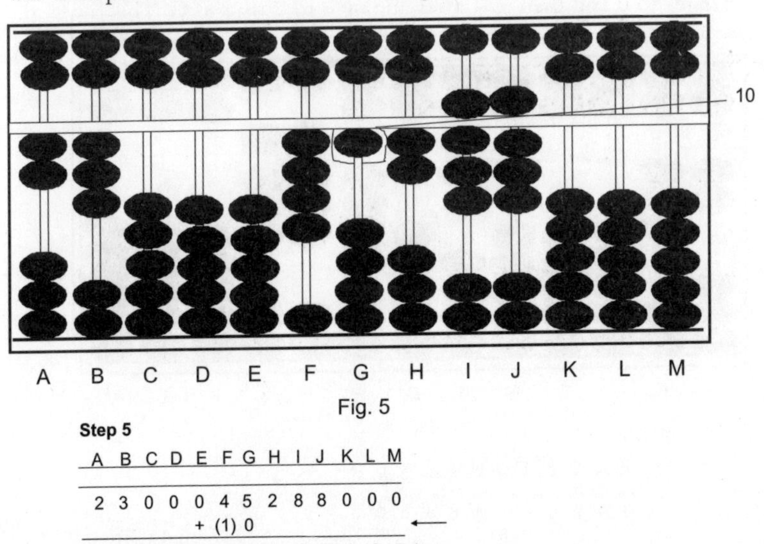

Fig. 5

Step 5

A	B	C	D	E	F	G	H	I	J	K	L	M	
2	3	0	0	0	4	5	2	8	8	0	0	0	
				+	(1)	0							←
2	3	0	0	0	4	1	2	8	8	0	0	0	←

Note: Difference in steps marked with ←

Step 6: Multiply 4 by 3 and add 12 to rods GH. (Fig. 6)

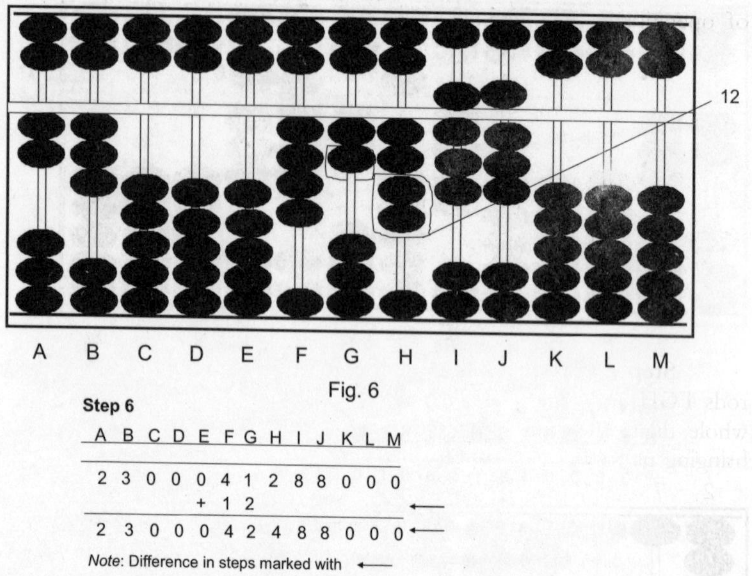

Fig. 6

Step 6

A	B	C	D	E	F	G	H	I	J	K	L	M
2	3	0	0	0	4	1	2	8	8	0	0	0
					+	1	2					
2	3	0	0	0	4	2	4	8	8	0	0	0

Note: Difference in steps marked with ←

Step 7 and the answer: Multiply 4 by 2 and add 08 to rods FG. Once again note the change; what was originally multiplicand 4 on F changes to 0 and then to 1 after adding 8 to 2 on rod G. This leaves the answer 10,488 on rods F through J. (Fig. 7)

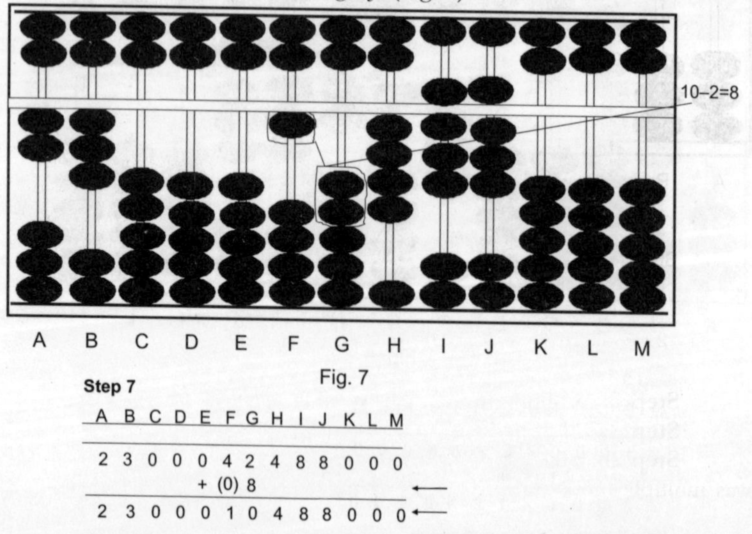

Fig. 7

Step 7

A	B	C	D	E	F	G	H	I	J	K	L	M
2	3	0	0	0	4	2	4	8	8	0	0	0
				+	(0)	8						
2	3	0	0	0	1	0	4	8	8	0	0	0

Note: Difference in steps marked with ←

Example 10: Prove that 347×276=95772 on an abacus.

The Order of Multiplication:

Please refer to Fig. 1 below to help with the illustration. The order of operation is as follows;

- Rod H×B×C×A.
  Note: the multiplicand on rod H changes to become part of the product.
- Rod G×B×C×A.
  Note: the multiplicand on rod G changes to become part of the product.
- Rod F×B×C×A.
  Note: the multiplicand on rod F changes to become part of the product.

Step 1: Choose *rod H* to be the unit rod. Set multiplicand 347 on rods FGH and multiplier 276 on rods ABC. The multiplier has three whole digits therefore the unit rod will shift three rods to the right bringing us to rod K.

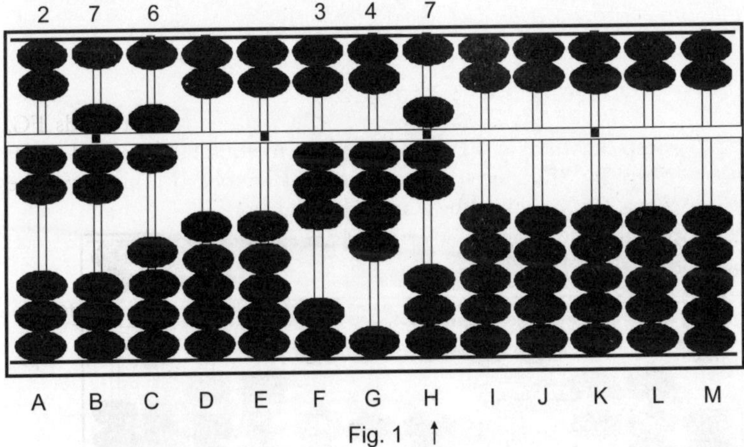

Fig. 1 ↑

Step 1

A	B	C	D	E	F	G	H	I	J	K	L	M
	.			.			.			.		
2	7	6	0	0	3	4	7	0	0	0	0	0

Step 2: Multiply *7 by 7 and add 49 to rods IJ.*

Step 2a: Multiply 7 by 6 and add 42 to rods JK.

Step 2b: Multiply 7 by 2 and add 14 to rods HI. Note that what was multiplicand 7 on H has now changed to 1. (Fig. 2)

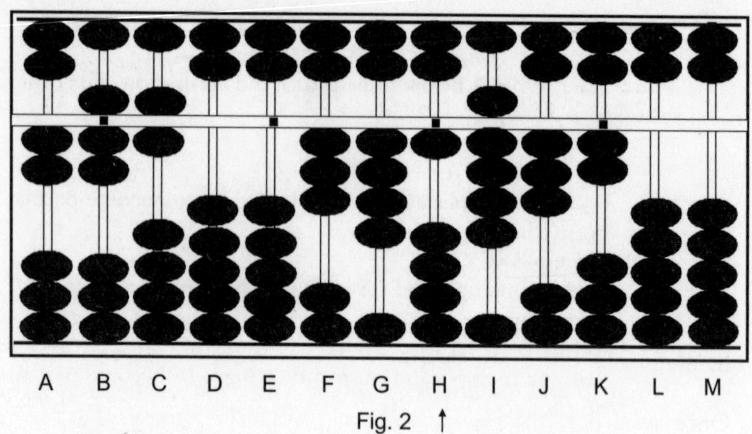

Fig. 2 ↑

Step 2

A	B	C	D	E	F	G	H	I	J	K	L	M
	·			·			·			·		
2	7	6	0	0	3	4	7	0	0	0	0	0
							+	4	9			
								+	4	2		
						+	(1)	4				
2	7	6	0	0	3	4	1	9	3	2	0	0

Step 3: Multiply 4 by 7 and add 28 to rods HI.

Step 3a: Multiply 4 by 6 and add 24 to rods IJ.

Step 3b: Multiply 4 by 2 and add 08 to rods GH. Note that what was originally multiplicand 4 on G changes to 0 and then to 1 after adding 8 to rod H. (Fig. 3)

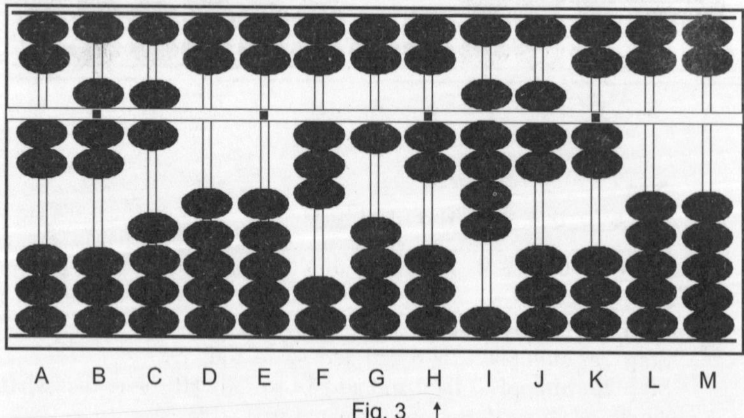

Fig. 3 ↑

Step 3

A	B	C	D	E	F	G	H	I	J	K	L	M
	.			.			.			.		
2	7	6	0	0	3	4	1	9	3	2	0	0
						+	2	8				
							+	2	4			
					+	(0)	8					
2	7	6	0	0	3	1	2	7	2	2	0	0

Step 4: Multiply 3 by 7 and add 21 to rods GH. 4a: Multiply 3 by 6 and add 18 to rods HI.

Step 4b and the answer: Multiply 3 by 2 and add 06 to rods FG. Once again note the change; what was originally multiplicand 3 on F changes to 0. This leaves the answer 95,772 on rods G through K. (Fig. 4)

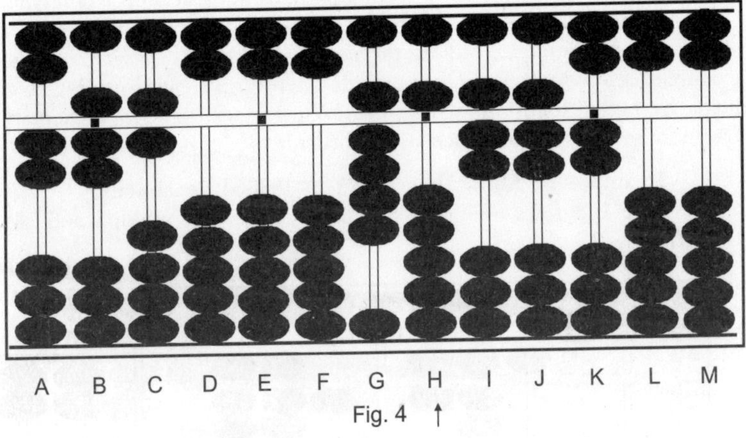

A B C D E F G H I J K L M

Fig. 4 ↑

Step 4

A	B	C	D	E	F	G	H	I	J	K	L	M
	.			.			.			.		
2	7	6	0	0	3	1	2	9	7	2	0	0
					+	2	1					
						+	1	8				
				+	(0)	6						
2	7	6	0	0	0	9	5	7	7	2	0	0

TRADITIONAL MULTIPLICATION TECHNIQUES

Chinese Suan Pan—The "Extra Bead" and the "Suspended Bead"—*Complementary Numbers*

Unlike the present day Japanese Soroban which has evolved to become an instrument with a 1/4 bead construction, the Suan Pan has retained its classic 2/5 bead construction—2 beads above the reckoning bar, 5 beads below. The techniques described below may offer some idea as to why this might be so.

- **The Extra Bead** technique uses the top most heaven bead and has a value of 5.
- **The Suspended Bead** technique also uses the top most heaven bead but in this case it's placed 1/2 way down the rod so that the bead neither touches the frame above nor the bead below; it has a value of 10.

Before beginning these techniques, it would be useful to have a good understanding of both the addition and multiplication techniques.

"The Extra Bead" When it comes to adding products onto the abacus, most techniques require the use of *complementa,y numbers*. In the examples below we'll use this approach wherever we can. In doing so it helps to better focus on where the exclusive use of complementary numbers makes the job much more difficult.

Example 11: Show that 189×576=108864 on abacus.

Step 1: Set the multiplier 576 on DEF and the multiplicand 189 on HIJ. (Fig. 1)

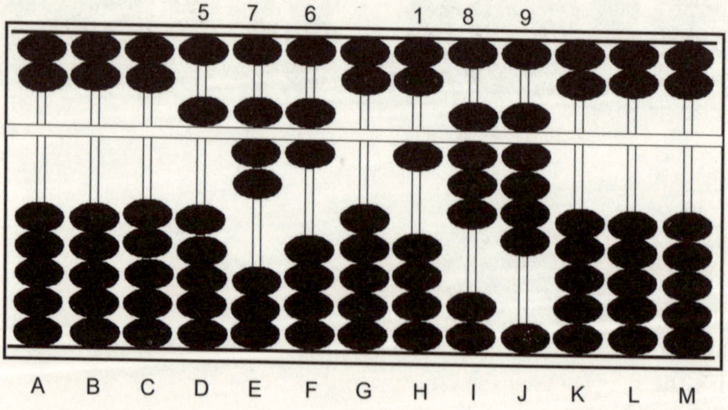

5 7 6 1 8 9

A B C D E F G H I J K L M

Fig. 1

Step 1

A	B	C	D	E	F	G	H	I	J	K	L	M
0	0	0	5	7	6	0	1	8	9	0	0	0

Step 2: Multiply the 9 on rod J by the 7 on E and add the product 63 on KL.

Step 2a: Multiply the 9 on rod J by the 6 on F and add the product 54 on LM

Step 2b: Multiply the same 9 on rod J by the 5 on rod D, remove the 9 on J and place the product 45 on JK. This leaves 185184 on rods HIJKLM. (Fig. 2)

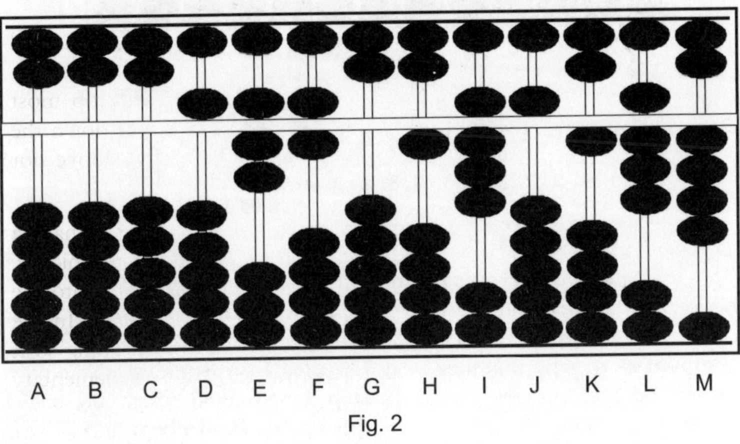

Fig. 2

Step 2

A	B	C	D	E	F	G	H	I	J	K	L	M	
0	0	0	5	7	6	0	1	8	9	0	0	0	
									+	6	3		Step 1
										+	5	4	Step 2a
									(4)	5			Step 2b
0	0	0	5	7	6	0	1	8	5	1	8	4	

This is where we use the *Extra Bead*. At this point we're going to multiply the 8 on rod I by the 7 on E. If we were to do a carry here then we'd have to change the 8 to 9. That would be confusing. (Especially since it's the 8 that's going to be multiplied by each of the digits in the multiplier.) Since there is an extra 5 bead available on J we'll use it to avoid confusion.

Step 3: Multiply the 8 on rod I by the 7 on E and add the product 56 on JK, leaving 1 8 "10" 7 8 4 on rods HIJKLM. (Fig. 3)

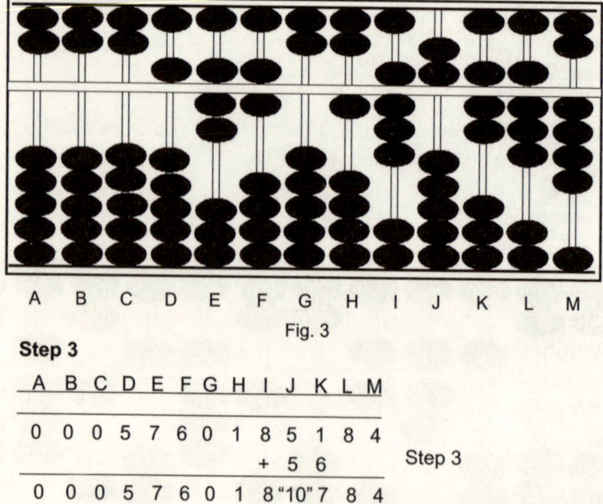

Fig. 3

Step 3

A	B	C	D	E	F	G	H	I	J	K	L	M
0	0	0	5	7	6	0	1	8	5	1	8	4
								+	5	6		
0	0	0	5	7	6	0	1	8	"10"	7	8	4

Step 3

It may be interesting to note that many Japanese Soroban made between circa 1850 and 1930 have one heaven bead and the earth beads. (i.e. I bead above the reckoning bar and 5 beads below.) we always been confused as to why this might be. Learning from some of these older techniques may provide a clue. In step 3 one could, as an alternative, show the number 10 on rod J as one heaven bead down and 5 earth beads up.

Step 3a: From here we can continue. Multiply the 8 on rod I by the 6 on F and add the product 48 on KL

Step 3b: Multiply the 8 on rod I by the Son 0, remove the 8 on land place the product 40 on rods IJ. This leaves 1 4 "11" 2 6 4 on rods HUKLM.

Step 3c: Next clear the 10 from rod J and carry it over to rod I changing the 4 to 5. This leaves 151264 on rods HIJKLM. (Fig. 4)

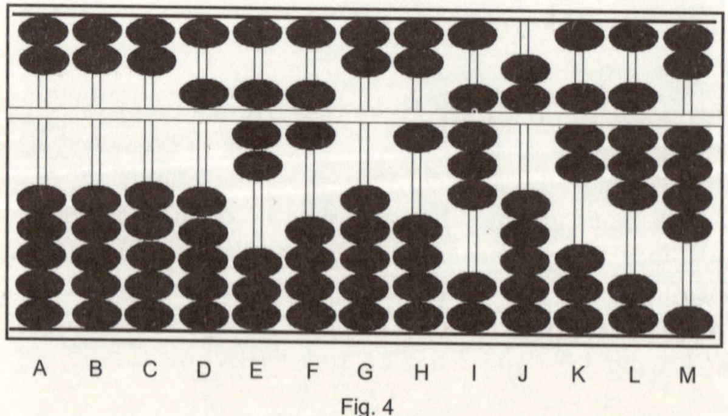

Fig. 4

Steps 3a-3b-3c

A	B	C	D	E	F	G	H	I	J	K	L	M	
0	0	0	5	7	6	0	1	8	"10"	7	8	4	
									+	4	8		Step 3a
							+	(4)	0				Step 3b
0	0	0	5	7	6	0	1	4	"11"	2	6	4	
+ extra bead								4	10				Step 3c
0	0	0	5	7	6	0	1	5	1	2	6	4	

Step 4: Now we can continue using complements in the normal way. Multiply the 1 on rod H by the 7 on rod E and add the product 7 on rod J.

Step 4a: Multiply the 1 on rod H by the 6 on rod F and add the product 6 on rod K.

Step 4b and the answer: Multiply the same 1 on rod H by the Son rod D, remove the 1 on rod H and add the product 5. This leaves 108,864 on rods HUKLM, which is the answer. (Fig. 5)

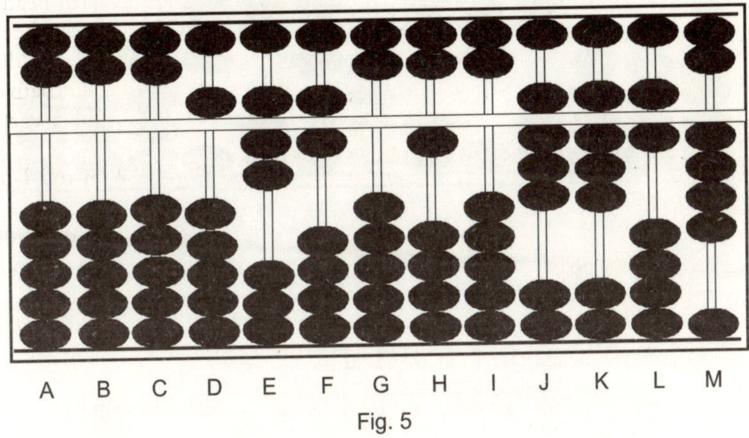

Fig. 5

Step 4

A	B	C	D	E	F	G	H	I	J	K	L	M	
0	0	0	5	7	6	0	1	5	1	2	6	4	
							+	0	7				Step 4
								+	0	6			Step 4a
							+	0	5				Step 4b
0	0	0	5	7	6	0	1	0	8	8	6	4	

"The Suspended Bead" (Xuan-zhu)

In this next example we'll use a technique where we bring down the upper most 5 bead A way down the rod so that it touches neither the frame above nor the bead below. This Suspended Bead has a value of 10. Once again in solving this next example, we'll use complementary numbers as much as possible. However, in two instances it will be clear that having extra beads simplifies the process.

Example 12: 989×898=888122

Step 1: Set the multiplier 898 on DEF and the multiplicand 989 on HIJ. (Fig. 1)

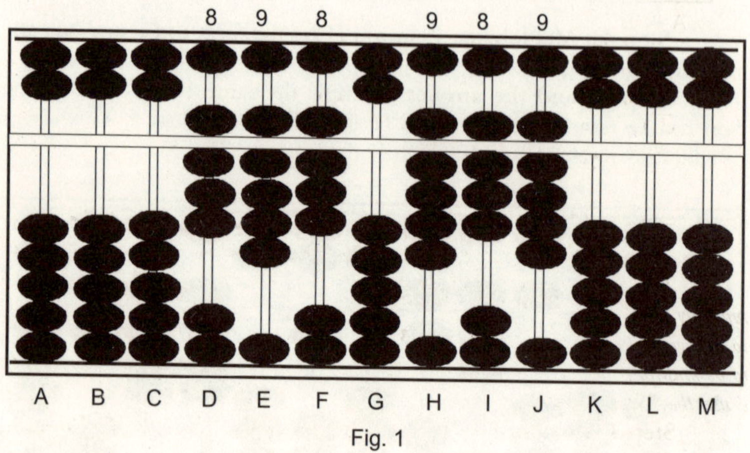

Fig. 1

Step 1

A	B	C	D	E	F	G	H	I	J	K	L	M
0	0	0	8	9	8	0	9	8	9	0	0	0

Step 2: Multiply the 9 on rod J by the 9 on E and add the product 81 on KL.

Step 2a: Multiply the 9 on rod J by the 8 on F and add the product 72 on LM.

Step 2b: Multiply the same 9 on rod J by the 8 on rod D, remove the 9 on J and place product 72 on JK. This leaves 988082 on rods HIJKLM. (Fig. 2)

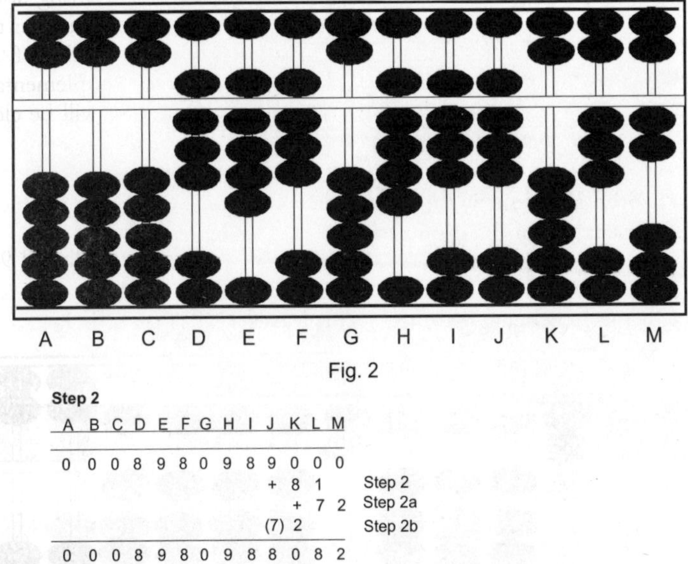

Fig. 2

Step 2

A	B	C	D	E	F	G	H	I	J	K	L	M	
0	0	0	8	9	8	0	9	8	9	0	0	0	
									+	8	1		Step 2
										+	7	2	Step 2a
									(7)	2			Step 2b
0	0	0	8	9	8	0	9	8	8	0	8	2	

This is where we use the Suspended Bead. At this point we're going to multiply the 8 on rod I by the 9 on E. If we were to do a carry here then we'd have to change the 8 to 9. That would be confusing. (Especially since it's the 8 that's going to be multiplied by each of the digits in the multiplier.) We might loose track. Instead e use the "Suspended Bead" technique.

Step 3: Multiply the 8 on rod I by the 9 on E and add the product 72 on JK. In bringing down the suspended bead to signify 10 and subtracting 3 we've added 7 to rod J giving the rod a total value of 15.

Step 3a: Multiply the 8 on rod I by the 8 on F and add the product 64 on KL. This leaves a value of 9 8 "15" 9 2 2 on rods HIJKLM. (Fig. 3)

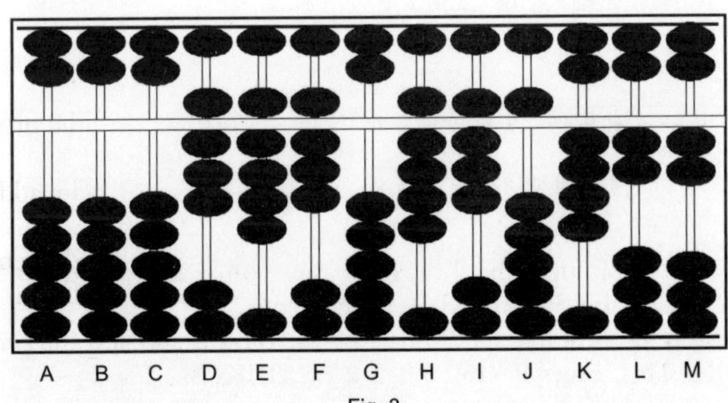

Fig. 3

Steps 3-3a

A	B	C	D	E	F	G	H	I	J	K	L	M	
0	0	0	8	9	8	0	9	8	8	0	8	2	
								+	7	2			Step 3
									+	6	4		Step 3a
0	0	0	8	9	8	0	9	8	"15"	9	2	2	

Step 3b: Multiply the same 8 on rod I by the 8 on rod D, remove the 8 on land place the product 64 on IJ. This leaves 9 6 "19" 9 2 2 on rods on rods HIJKLM.

Step 3c: Next clear the "Suspended" 10 from rod J and carry it over to rod I changing the 6 to 7. This leaves 979922 on rods HIJKLM. (Fig. 4)

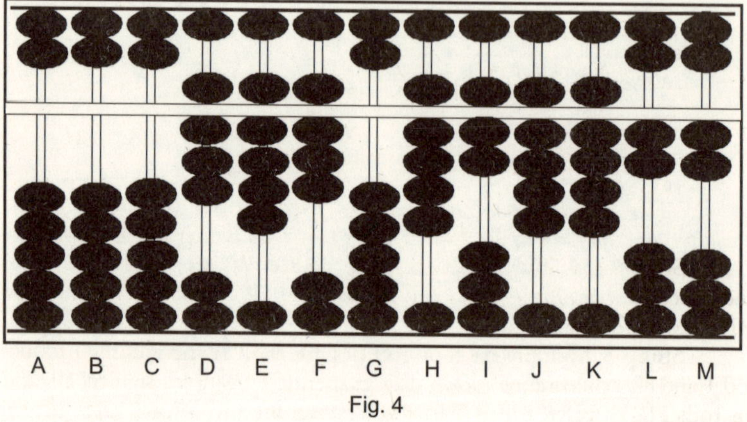

Fig. 4

Steps 3b-3c

A	B	C	D	E	F	G	H	I	J	K	L	M	
0	0	0	8	9	8	0	9	8	"15"	9	2	2	
								(6)	4				
									+	6	4		Step 3b
0	0	0	8	9	8	0	9	6	"15"	9	2	2	
+	suspended							1	0				Step 3c
0	0	0	8	9	8	0	9	7	9	9	2	2	

Step 4: Multiply the 9 on H by the 9 on E and add the product 81 on IJ using the Suspended Bead technique.

Step 4a: Multiply the 9 on H by the 8 on F and add the product 72 on JK. This leaves 9 "16" 8 1 2 2 on HIJKLM. (Fig. 5)

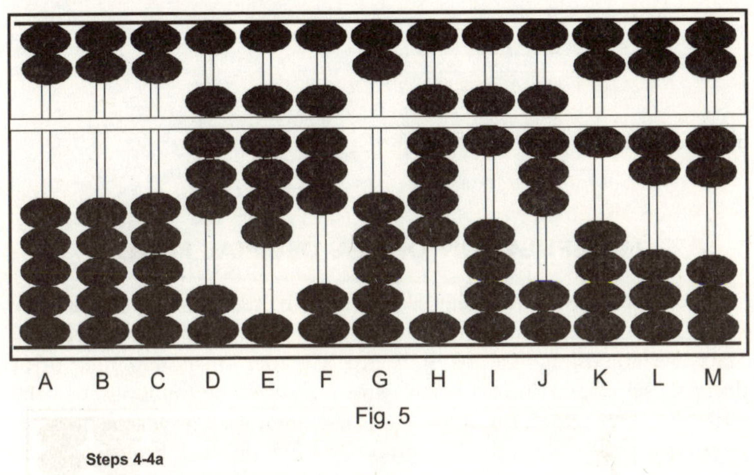

Fig. 5

Steps 4-4a

A	B	C	D	E	F	G	H	I	J	K	L	M	
0	0	0	8	9	8	0	9	7	9	9	2	2	
							+	8	1				Step 4
								+ 7	2				Step 4a
0	0	0	8	9	8	0	9	"16"	8	1	2	2	

Step 4b: Multiply the 9 on H by the 8 on D, remove the 9 on H and place the product 72 on HI. This leaves 7 "18" 8 1 22 on rods HIJKLM.

Step 4c and the answer: Next clear the "Suspended" 10 from rod I and carry it over to rod H changing the 7 to 8. This leaves 888122 on rods HIJKLM, which is the answer. (Fig. 6)

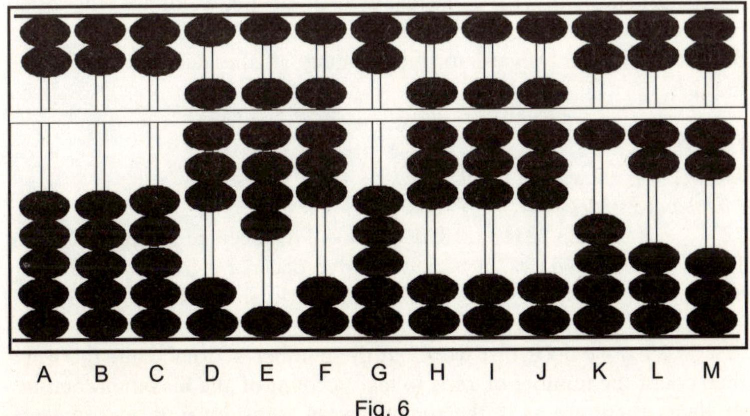

Fig. 6

Steps 4b-4c

A	B	C	D	E	F	G	H	I	J	K	L	M	
0	0	0	8	9	8	0	9	"16"	8	1	2	2	
								(7) 2					Step 4b
0	0	0	8	9	8	0	7	"18"	8	1	2	2	
								missing					Step 4c
0	0	0	8	9	8	0	7	"18"	8	1	2	2	

MANIPULATION OF THE DECIMAL POINT

We present here a simple method for manipulating the decimal point which basically stems from arithmetic, rather from abacus tradition. Since we are using base 10, it is just a matter of noting how many decimal significant places there are in both the multiplicand and multiplier. Our reason for doing this is itself simple. The abacus does not physically dictate where the last integer and first decimal digits are. At the end of the day, the calculations produce a bead representation of the final product or result. All that is needed if there were decimals involved would be ascertaining where the decimal point should be placed.

A decimal number contains three parts, the integer part (I), the decimal point (.), and finally the decimal part (D) or I.D. Supposing we had a number which can be represented by III.DDDDD (e.g. 123.12345).

We see that the integer part contains three digits. The decimal part contains five digits When placing the number onto the abacus, we cannot indicate the decimal point in anyway. We would have just loaded in 12312345 as in the example above.

Integer Multiplied by a Decimal

When multiplying a decimal with an integer, we notice what often happens is that there will be the same number of decimal places to the decimal significant figure in the product as the decimal multiplying number.

Example:

Integer Multiplicand	*Decimal Multiplier*	*Decimal Product*
723	×123.12345	=89018.25435 Five decimal significant places
12	×1.333	=15.996 Three decimal significant places
817	×1.1	=898.7 One decimal significant place

What we do is first work out the number of total digits there are, then count the number of rods to take account of the maximum number of digits there are as if the multiplication were done on two integers.

Once the product is obtained, the number of decimal significant places is counted from the right of the abacus and you have the answer you seek.

Decimal Multiplied by another Decimal

When two decimals multiply, the number of decimal places to the decimal significant figure is indicated by how many decimal digits in total there are in both numbers.

Example:

Decimal Multiplicand	*Decimal Multiplier*	*Decimal Product*
723.3	×123.12345	=89055.191385 Six decimal significant places
12.217	×1.333	=16.285261 Six decimal significant places
817.12378	×1.1	=898.836158 Six decimal significant place

A similar process is used; counting all the digits and making space on the abacus, obtaining the product before noting how many decimal significant places in the original multipliers (the multiplicand and multiplier) before counting from the right of the abacus to find the correct positioning of the decimal point on the abacus representation.

Example 13: Evaluate 27.42×3.91

The problem is to solve first **2742×391**

Consider these two numbers as integers.

There are seven digits in total, four in 2742 and three in 391

```
2742 391***   Abacus is loaded thus Abacus reads 391****
                                                  ++
   *2      The product of 1 and 2 Abacus reads 3902**
    *7     The product of 1 and 7 Abacus reads 39027**
     *4    The product of 1 and 4 Abacus reads 390274*
      *2   The product of I and 2  Abacus reads 3902742
                                                  *
  18       The product of 9 and 2  Abacus reads 3182742
   63      The product of 9 and 7  Abacus reads 3245742
    36     The product of 9 and 4  Abacus reads 3249342
     18    The product of 9 and 2  Abacus reads 3249522
   *6      The product of 3 and 2  Abacus reads 0849522
    21     The product of 3 and 7  Abacus reads 1059522
     12    The product of 3 and 4  Abacus reads 1071522
      *6   The product of 3 and 2  Abacus reads 1072122
```

The answer is the product **1072122**. However, we must now put back in the number of decimal significant places in our original problem (27.42×3.91) We note that there are four decimal significant places in total, therefore, we must count four digits from the right to obtain the correct answer: **107.212=27.42×3.91.**

There is one special case worth mentioning. That is when one or both numbers has zero as the integer part of the decimal. In this case, use only the significant numbers in the multiplication, and then follow the method for finding the correct position of the decimal point as usual.

0.028×0.11
28×11=308

There are five decimal significant places, therefore the answer is 0.00306 0.028×0.11 as the correct answer. That's all there is to decimal multiplication.

Division Techniques

The Chinese division technique works best with problems where divisors are made up of whole numbers and have a value of less than 10.

Now let us try to do something challenging on the abacus:

Division

An abacus division is tricker than abacus multiplication. But the technique that is used to do division on the abacus is an important fundamental one; it's what makes it possible to use the abacus for more advanced operations, like roots.

Below is a list of traditional rules. Understanding these rules is integral to understanding this method. They may look a little complicated at first glance but they are easily understood. As such there should be no need of conmit them to memory. (See below for a brief explanation.)

Traditional Rules

1/1 forward 1		1/8=1 plus 2
		2/8=2 plus 4
1/2=5	1/6=1+4	3/8=3 plus 6
2/2=forward 1	2/6=3 plus 2	4/8=5
	3/6=5	5/8=6 plus 2
1/3=3 plus 1	4/6=6 plus 4	6/8=7 plus 4
2/2=6 plus 2	5/6=8 plus 2	7/8=7 plus 6
3/2=forward 1	6/6=forward 1	8/8=forward 1

1/4=2 plus 2
2/4=5
3/4=7 plus 2
4/4=forward 1

1/5=2
2/5=4
3/5=6
4/5=8
5/5=forward 1

1/7=1 plus 3
2/7=2 plus 6
3/7=4 plus 2
4/7=5 plus 5
5/7=7 plus 1
6/7=8 plus 4
7/7=forward 1

1/9=1 plus 1
2/9=2 plus 2
3/9=3 plus 3
4/9=4 plus 4
5/9=5 plus 5
6/9=6 plus 6
7/9=7 plus 7
8/9=8 plus 8
9/9=forward 1

In the above set of rules:

"plus 1 means adding 1 to the column immediately to the right, plus 2 means adding 2 and so on...

"forward 1" means adding 1 to the column immediately to the left.

The DIVIDEND is the number which is to be divided. The DIVISOR is the number which dives the dividend.

DETERMINATION OF THE UNIT POSITION FOR THE QUOTIENT:

(1) Count the figures in the integral part of the divisor. Count off an equal number of uprights, from the unit position of the dividend towards the left: count off one more; and the upright is the unit position for the quotient.

(2) In such case as there are some 0's (zero's) between the decil point and the decimal-significant figure of the divisor.

(a) Count the number of 0's ("zero"s) between the decimal point an the decimal—significant figure of the divisor. Count off an equal number of upright, from the unit position of th dividend towards the right: count off one less: and the last upright is the unit position for the quotient.

(b) In such case as there is one "0" ("Zero") between the decimal point and the decimal significant figure of the divisor, the unit position of the dividend is the unit position for the quotient.

(3) In such case as there is no "0" ("zero") between the decimal point and the decimal-significant figure, next upright from the unit position of the dividend towards the left is the unit position for the quotient.

Explahation of Rules

All is best explained using examples.

Using a few random examples, this is how I think of it:

(a) 2/3=6 plus 2 ⇒20 ÷3=6 with remainder 2
Place 2 one column to the right

(b) 6/7=6 plus 4⇒60 ÷7=8 with remainder 4
Place 4 one column to the right

(c) 4/7=5 plus 5 ⇒40 ÷7=5 with remainder 5
Place 5 one column to the right

(d) 5/5=forward 1⇒5 ÷5=1
Place 1 one column to the left...
and so on

Example 1: Show that 128÷2=64 on abacus.

Step 1: Set the dividend 128 on FGH and the divisor 2 on C. (Fig. 1(a))

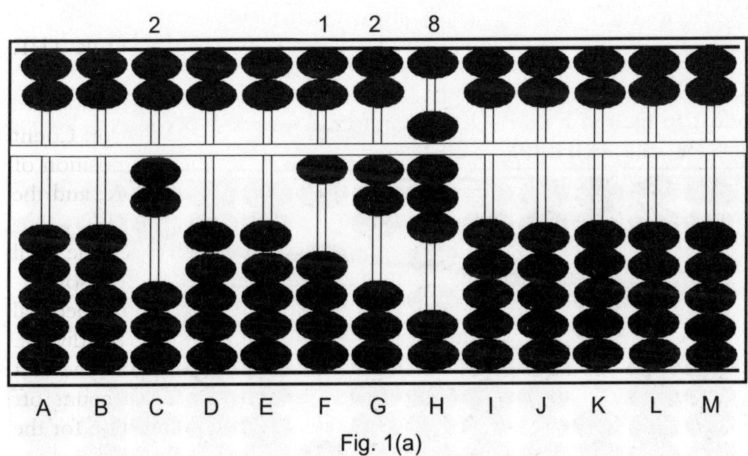

Fig. 1(a)

Step 1

A	B	C	D	E	F	G	H	I	J	K	L	M	
0	0	2	0	0	1	2	8	0	0	0	0	0	Step 1

Step 2: Following rule 1/2=5⇒ divide 2 on rod C into 1 on F. (Think of it in terms of 10÷2=5). Remove the 1 and replace it with the quotient 5 on rod F, leaving 528 on rods FGH. (Fig. 1(b))

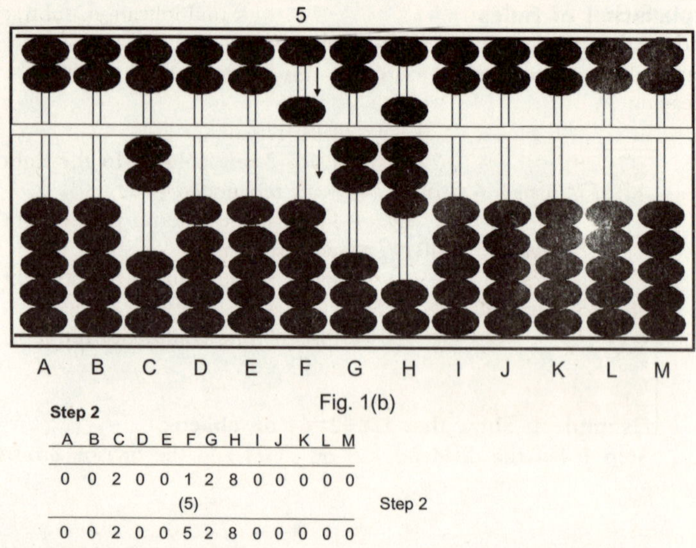

Fig. 1(b)

Step 2

	A	B	C	D	E	F	G	H	I	J	K	L	M	
	0	0	2	0	0	1	2	8	0	0	0	0	0	
						(5)								Step 2
	0	0	2	0	0	5	2	8	0	0	0	0	0	

Step 3: Now consider 2 as divisor of rod C and 2 of 528 on rod G. Following rule 2/2=forword 1⇒ this means clear the 2 on rod G and add 1 to the rod F on the left. Therefore, 5 on rod F becomes 6, leaving 608 on rods FGH. (Fig. 1(c))

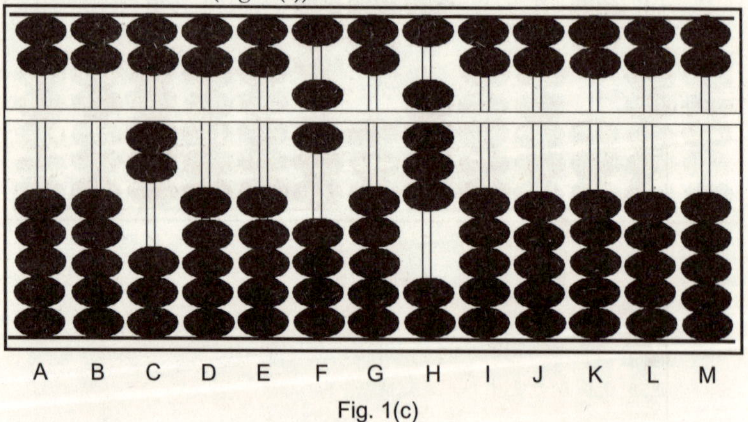

Fig. 1(c)

Step 3

	A	B	C	D	E	F	G	H	I	J	K	L	M	
	0	0	2	0	0	5	2	8	0	0	0	0	0	
clear							(2)							
forward						1								Step 3
	0	0	2	0	0	6	0	8	0	0	0	0	0	

Step 4: Lastly, consider 2 as divisor on rod C and 8 on rod H of dividend. Following rule 8/2=forward 4 ⇒ this means clear the 8 on rod H and add 4 to the rod G on the left. Therefore, 0 on rod G becomes 4, leaving 64 on rods FG, which is the answer. (Fig. 1(d))

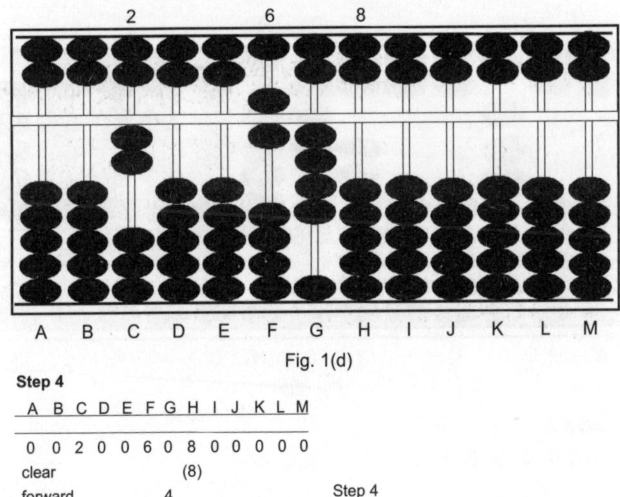

Fig. 1(d)

Step 4

	A	B	C	D	E	F	G	H	I	J	K	L	M	
	0	0	2	0	0	6	0	8	0	0	0	0	0	
clear								(8)						
forward							4							Step 4
	0	0	2	0	0	6	4	0	0	0	0	0	0	

Thus, 128÷2=64.

Example 2: Show that 259÷7=37

Step 1: Set the dividend 259 on FGH and the divisor 7 on C. (Fig. 2(a))

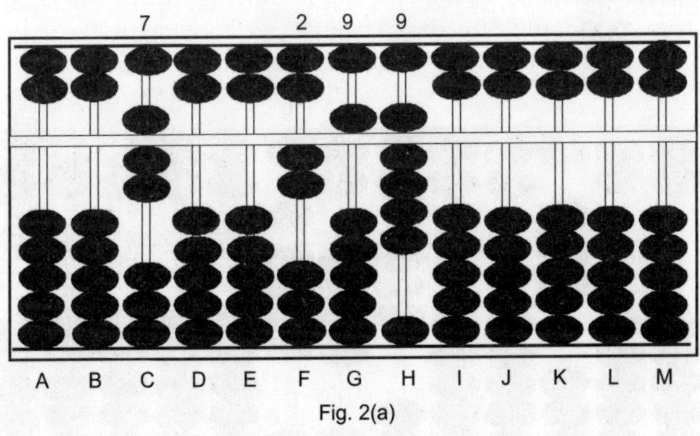

Fig. 2(a)

Step 1

A	B	C	D	E	F	G	H	I	J	K	L	M	
0	0	7	0	0	2	5	9	0	0	0	0	0	Step 1

Step 2: Following rule 2/7=2 plus 6 ⇒ divide 7 on rod C into 2 on F. (Think of it in terms of 20÷7=2 with a remainder of 6). Place the quotient 2 on rod F and place the remainder 6 on G, leaving 2 "11" 9 on rods FGH. (Fig. 2(b))

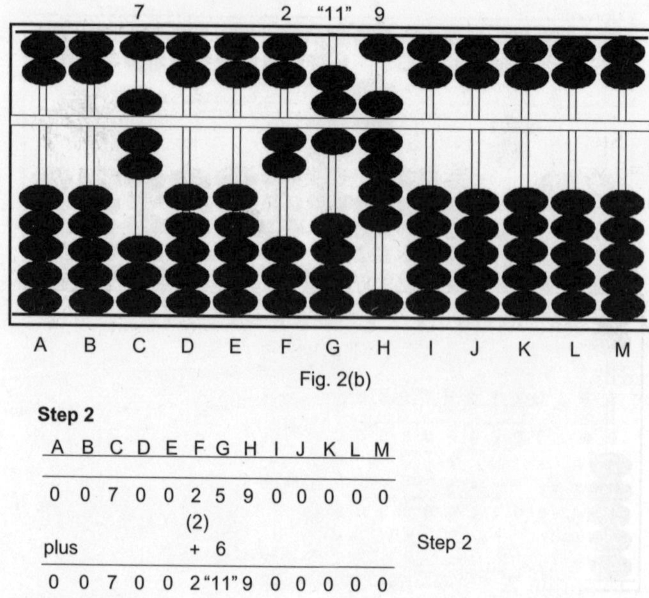

Fig. 2(b)

Step 3: The 11 on rod G is a number greater than 7. So take 7 away from rod G and *forward 1* onto rod F changing the 2 to a 3and leave 4 beads on G as 11–7=4. This leaves 349 on rods FGH. (Fig. 2(c))

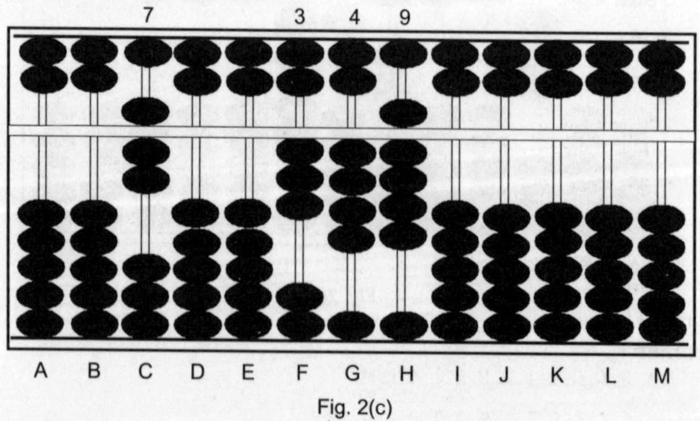

Fig. 2(c)

Step 3

A	B	C	D	E	F	G	H	I	J	K	L	M
0	0	7	0	0	2	"11"	9	0	0	0	0	0
						− 7						
forward					1							
0	0	7	0	0	3	4	9	0	0	0	0	0

Step 3

Step 4: Following rule 4/7=5 plus 5 ⇒ (Think of it in term of 40÷7=5 with a remainder of 5). Replace the 4 on G with the quotient 5. Place the remainder 5 onto rod H, leaving 3 5 "14" on rods FGH. (Fig. 2(d))

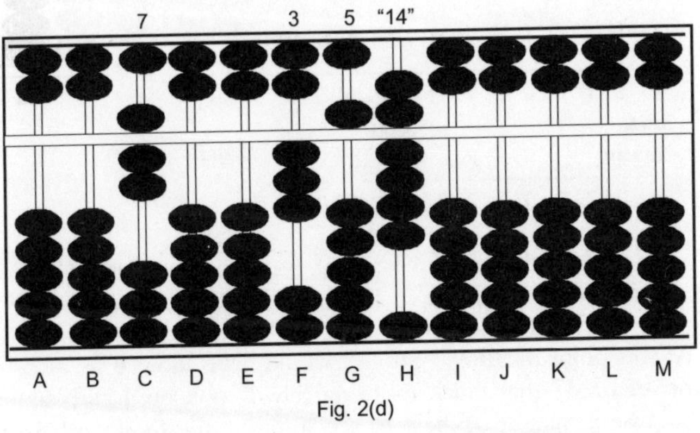

Fig. 2(d)

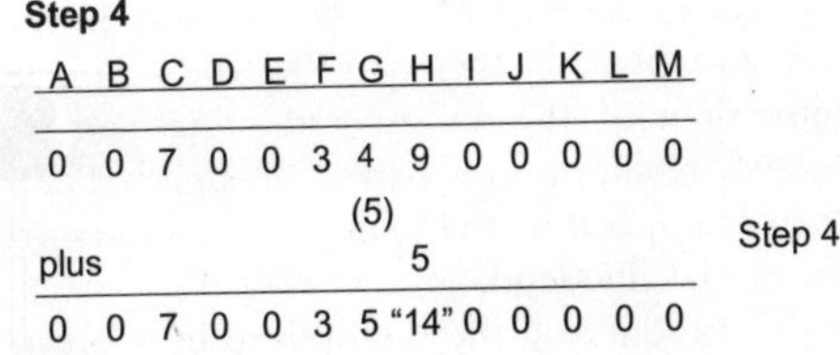

Step 4

A	B	C	D	E	F	G	H	I	J	K	L	M
0	0	7	0	0	3	4	9	0	0	0	0	0
						(5)						
plus							5					
0	0	7	0	0	3	5	"14"	0	0	0	0	0

Step 4

Step 5: Following rule 14/7=forward 2 ⇒ this means clear the 14 on rod H and add 2 to the rod G on the left. Therefore, 5 on rod G becomes 7, leaving 37 on rods FG, which is the answer. (Fig. 2(e))

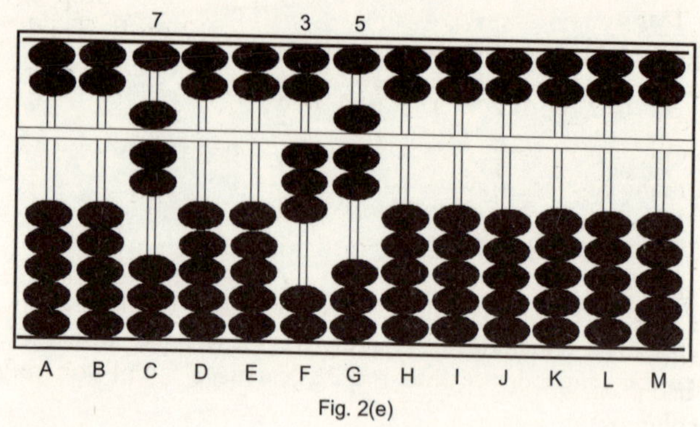

Fig. 2(e)

Step 5

	A	B	C	D	E	F	G	H	I	J	K	L	M	
	0	0	7	0	0	3	5	"14"	0	0	0	0	0	
clear								"14"						
forward							2							Step 5
	0	0	7	0	0	3	7	0	0	0	0	0	0	

It's evident that some of the above steps could have been done more efficiently. For instance in Example 2: Step 3, instead of dividing 7 into the 4 on rod G, we could have looked ahead and noticed 49 on GH. At this point dividing 7 into 49 would have allowed us to come to our answer of 37 that much more quickly. I took the longer approach only because it may be of help to see some of the thinking behind the process.

The initial digit of the dividend (placed near the middle of the abacus) is compared with the initial digit of the divisor and the multiples are noted on the abacus someplace in the left most rods. This multiple is multiplied by the initial digit of the divisor and subtracted from the initial digit of the dividend, resulting in a partial dividend. The second digit of the divisor is then multiplied by the multiple and subtracted from the second digit of the partial dividend.

(It is likely that this product of the multiplication is greater than the number represented by the first and second rods. If so, the initial multiple is reduced in value by one, and this new multiple multiplied by the initial dig of the divisor is added back to the initial digit of the dividend, mind this forms our first revised partial dividend. This way, the revised multiple multiplied by the second digit of the divisor may be small enough to be subtracted from the left most rods of the dividend. All will become apparent in the example below) .

This same process continues for each digit of the divisor until the last digit for the first multiple. If there are any remaining digits to the dividend, the whole process is repeated for the second digit of the multiple.

Once there are no digits left in the dividend, the answer will be all the multiple digits found.

Example 3: Let's divide 4582 by 17 on abacus.

With the abacus, we start by setting the reference column. The way that we do that is by deciding *how many digits* we want in our answer. For our example, we can say three digits—because we can easily see that the integer part of dividing 4582 by 17 will be three digits (even if we hadn't already worked out the answer). So, we'll use the leftmost three columns for building the quotient. We zero out the quotient area, and put out divisor into the rightmost columns. The initial setup looks like Figure (a) below.

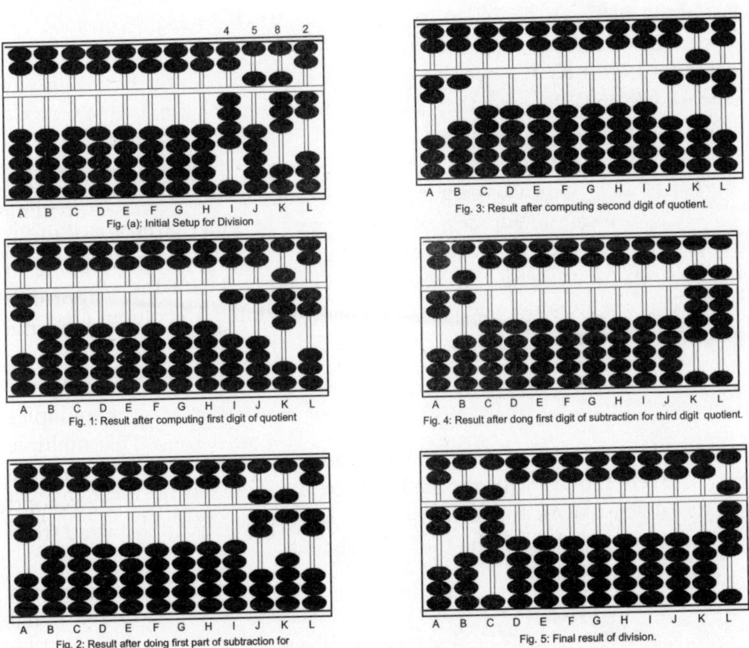

Fig. (a): Initial Setup for Division

Fig. 1: Result after computing first digit of quotient

Fig. 2: Result after doing first part of subtraction for second digit of quotient.

Fig. 3: Result after computing second digit of quotient.

Fig. 4: Result after dong first digit of subtraction for third digit quotient.

Fig. 5: Final result of division.

With the abacus set up, start following the same basic algorithm that is used on paper. What's the first digit of the quotient? It's two. So put a 2 in the leftmost column A of the quotient area; and then start to multiply 2×17; but instead of adding the results as done in multiplication, subtract them from the dividend, i.e. 17×2=34 subtract 3 from the column rod I and 4 from the column J (5–4=1; so one bead is left or rod J). The abacus now looks like Figure 1 in the image.

Now, look at the dividend section of the abacus, and find the rightmost column where the digits up to that column are larger than 17. That's one column to the right of where we started working last time: the second column from the right. So the working digits are 118. The largest multiple of 17 smaller than 118 is 6×17. Therefore, subtract 6×7=42 from columns K and J, which requires borrowing 1 from column I. So after this part of the subtraction, the abacus looks like Figure 2. Now subtract 6×1=6 from column J. Finally, add this new digit of the quotient to the second column from the left of the quotient area. The abacus now looks like Figure 3.

Finally, for the third digit of the quotient. The working remainder is the three remaining digits (because the working remainder from the last digit is 16, we shift right, and out working remainder becomes the last three digits, 162.) 162 is very-close to 10×17, thus, 9 must be the next digit of the quotient. So, multiply 9×7=63, and subtract that from columns L and K. We need to borrow one from column K to subtract 3 from column one; and then borrow one from column J to subtract 6 from column J. The result of that is shown in Fig 4. Then subtract 9×1=9 from column L; and put 9 in column J of the quotient. So the answer is 269, with a remainder of 9, as shown in Fig 5.

If you can do addition on paper, you can do it on the abacus, only faster. The real trick to doing complicated things is exactly what we saw in division; partitioning the abacus into regions. That's the reason why when you see a Chinese abacus, it's usually so long: 13 column abacuses are very common.

Similarly, division of two numbers x and y can be written as follows:

x/y

Here x is called the dividend and y is called the divisor.

Step 1: Start with the number for the dividend on the right half of the abacus. Work from left to right. Record the first digit (from left) of the result on the left half of the abacus.

```
  59           69
6)373       69)3732
 - 36        -  30   ← 1
   28        -  15   ← 2          6     3
   13          582                4     4
 - 12        -  54   ← 3            2 3
    1        -  27   ← 4        * 5     9
                15
```

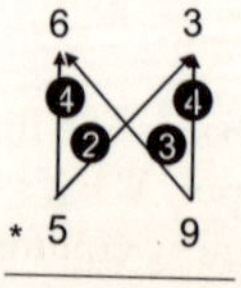

Step 2: Multiply the first digit of the result and the first digit (from left) of the divisor and subtract the number from the dividend.

Step 3: Multiply the first digit of the result and the second digit (from left) of the divisor and subtract the number from the remainder. As one moves from left to right, the product number should be subtracted on the abacus one column to the right.

Step 4: Repeat step 3 until we reach the end of divisor.

Step 5: Repeat steps 2-3 for the second digit of the result and so on. As one moves from left to right for the result, the product number should also be subtracted on the abacus one column to the right.

Example 4: Evaluate 373/6

Step 1: Add 373 on the right half of the abacus. (Fig. (a))

Step 2: The first digit of the result from left is 6. Record this number on the left half of the abacus. (Fig. (b))

Step 3: Multiply the first digit from left for the result (6) and the first digit of the divisor (6) and subtract the number 36 (6×6) from the dividend. Note the alignment should be on the left for both the dividend (373) and 36. That is 360 is subtracted from 373. In some cases, when subtracting, align to the left most digit of the dividend may not be possible (e.g., 12/8) then the multiplication product should be aligned to the second digit from left for the dividend. (Fig. (b))

Step 4: The second digit from left for the result is 2. Record this number next to the first digit already on the abacus.

Step 5: Multiply 2 and the divisor, and subtract the result (12) from the remainder number on the abacus. Note when subtracting, we need to shift one column to the right since we have shifted one column to the right for the result. So 12 are subtracted from the remainder.

Step 6: One may continue to get results after the decimal points. (Fig. (c))

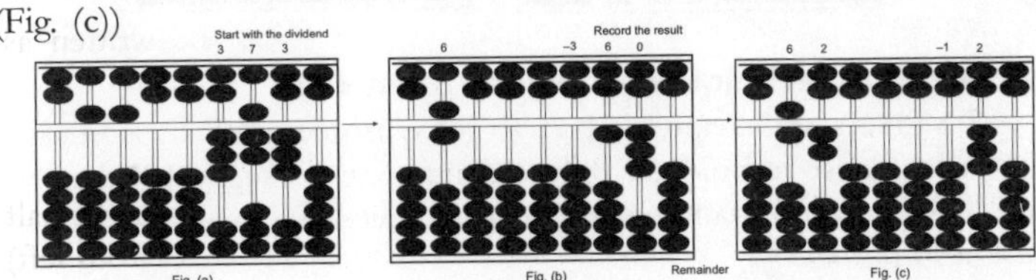

Fig. (a) Fig. (b) Fig. (c)

Example 5: Evaluate3732/63

Step 1: Add 3732 to the right half of the abacus. (Fig. (a))

Step 2: Start from left to right. Record the first digit of the result 5 on the left half of the abacus. (Fig. (b))

Step 3: Multiply 5 and the first digit (6) of the divisor, subtract the result 30 from the dividend.

Note: 30 should be aligned to the left of the dividend; that is we actually subtract 3000 in this example.

Step 4: Multiply 5 and the second digit (3) of the divisor, subtract the result 15 from the remainder on the abacus. Since we have shifted one column right for the divisor, we also need to shift one column right when subtracting. We actually subtract 150 in this example.

Step 5: Now we repeat for the second digit for the result. Record 9 next to 5 already on the abacus. (Fig. (c))

Step 6: Multiply 9 and the first digit (6) of the divisor, subtract the result 54 from the remainder. Note 54 should be aligned to the second digit from left of the dividend since we have shifted one column right for the result; that is we actually subtract 540 in this example. (Fig. (d))

Step 7: Multiply 9 and the second digit (3) of the divisor, subtract the result 27 from the remainder. We need to shift right one more column since we shifted one column right for the divisor. In the example, we subtract exactly 27. (Fig. (e))

Step 8: One may continue to get results after the decimal points.

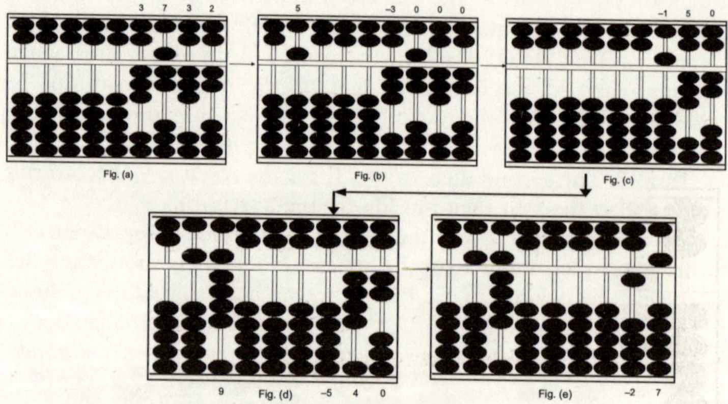

Long Division Techniques—Chinese Suan Pan

It's the second chapter in how to solve problems of division using traditional Chinese techniques. Before continuing it's suggested that one should have an understanding of the techniques and rules that govern *Chinese short division.*

This method of long division uses a traditional 2:5 bead suan pan. In order to solve the problem it's best to have an understanding of *Multiplication Techniques*;

1.The Extra Bead Technique which uses the top most heaven bead. It has a value of 5.

2.The Suspended Bead Technique which also uses the top most heaven bead but in this case it is set 1/2 way down the rod so that the bead neither touches the frame above nor the bead below. It has a value of 10.

It Should be noted that as with any method for solving problems of division or, an abacus, there will be instances where the abacus operator will have to revise a quotient answer either up or down in order to come to the correct result. Although remarkably good, this technique is no exception.

Below are examples for this technique. Even though the first two examples 6 and 7 have decimal answers they are quite straight forward and demonstrate the method very well. For our purposes it will be enough to solve each of these problems to four decimal places with a remainder at the end.

Example 6: Show that 290÷47=6.1702 ⇒ with remainder 0.0006 (see Fig. 1)

This example uses the rules;

(a) rule 2/4=5
(b) rule 4/4=forward 1
(c) rule 3/4=7 plus 2
(d) rule 1/4=2 plus 2

Step 1: Set the dividend 290 on rods EFG and divisor 47 on AB.

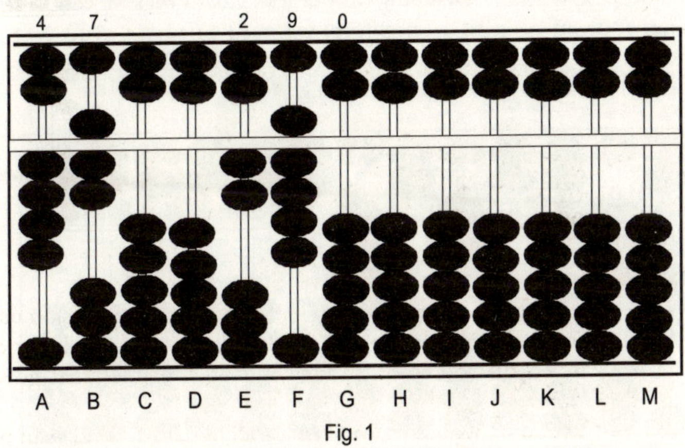

Fig. 1

Step 1

A	B	C	D	E	F	G	H	I	J	K	L	M	
4	7	0	0	2	9	0	0	0	0	0	0	0	Step 1

Step 2: Compare 4 on A and 2 on E. Follow rule: 2/4=5⇒ The first number in our quotient answer will be 5. Place 5 on rod E. Multiply 5 on E by 7 on B and subtract the product 35 from rods FG. (Fig. 2)

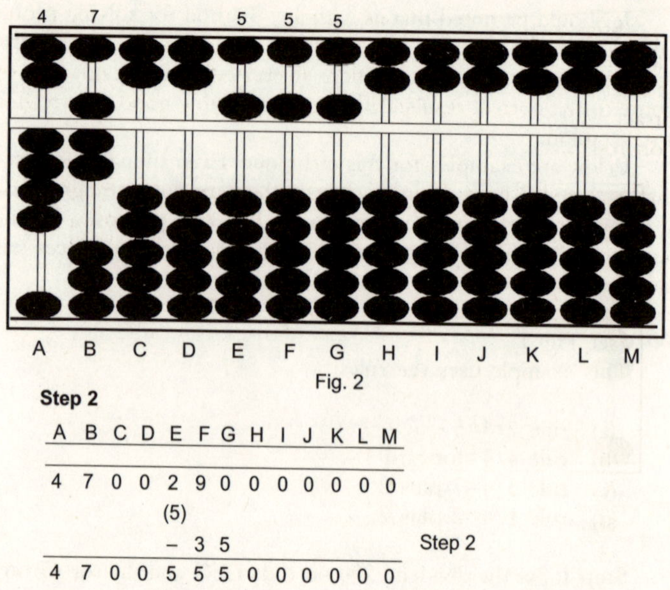

Fig. 2

Step 2

A	B	C	D	E	F	G	H	I	J	K	L	M	
4	7	0	0	2	9	0	0	0	0	0	0	0	
				(5)									
				–	3	5							Step 2
4	7	0	0	5	5	5	0	0	0	0	0	0	

Step 3: With 55 remaining on FG it's evident that we can subtract another 47 from rods FG. Therefore, we need to revise. Add 1 to the quotient on E and subtract a further 47 from rods FG. This leaves quotient 6 on E and remainder 8 on G. (Fig. 3)

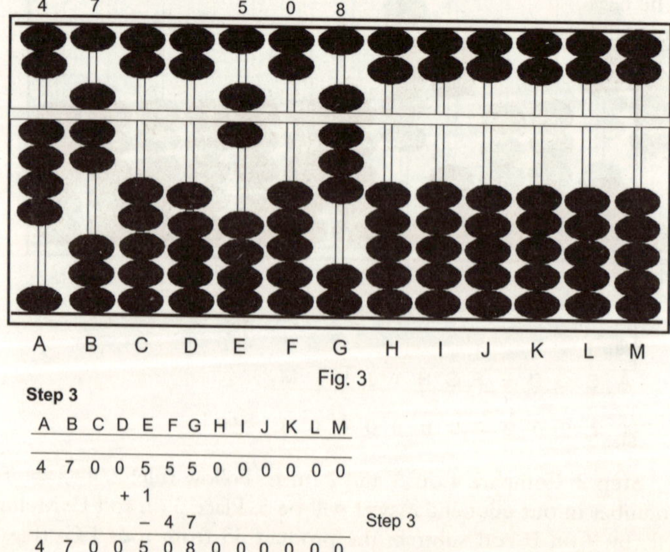

Fig. 3

Step 3

A	B	C	D	E	F	G	H	I	J	K	L	M	
4	7	0	0	5	5	5	0	0	0	0	0	0	
			+	1									
				–	4	7							Step 3
4	7	0	0	5	0	8	0	0	0	0	0	0	

Step 4: Compare 4 on A with 8 on G. If we forward 2 there won't be anything left to continue, therefore, follow rule: 4/4=forward 1 ⇒ The next number in the quotient answer will be 1. Forward 1 to F and subtract 47 from rods GH. This leaves quotient 61 on EF and remainder 33 on rods GH. (Fig. 4)

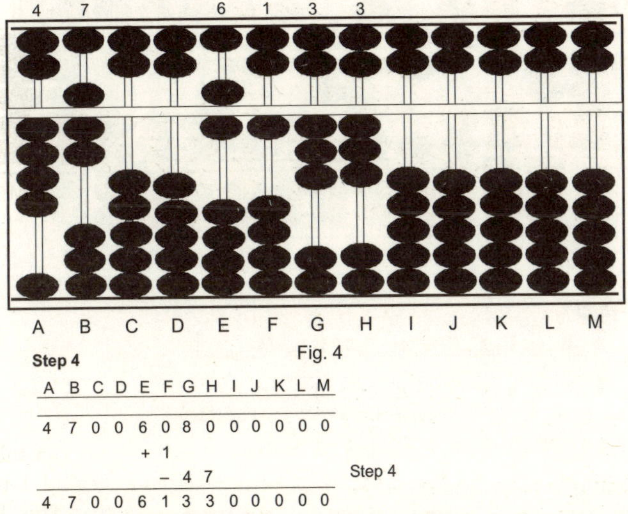

Fig. 4

Step 4

A	B	C	D	E	F	G	H	I	J	K	L	M	
4	7	0	0	6	0	8	0	0	0	0	0	0	
					+ 1								
						− 4	7						Step 4
4	7	0	0	6	1	3	3	0	0	0	0	0	

Step 5: Compare 4 on A and 3 on G. Follow rule: 3/4=7 plus 2 ⇒ The next number in our quotient will be 7. Place 7 on rod G and add the remainder 2 to rod H. (Fig. 5)

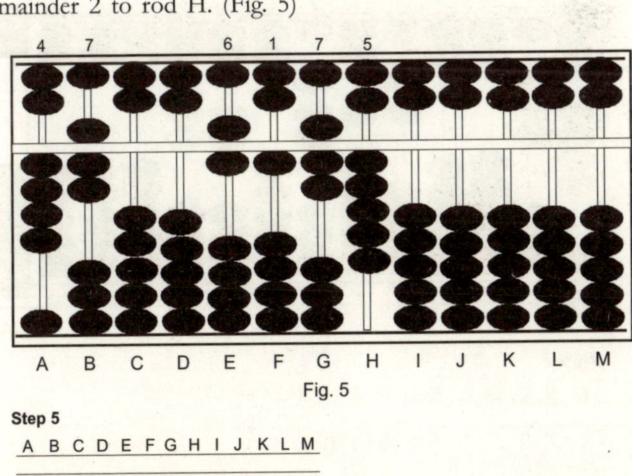

Fig. 5

Step 5

A	B	C	D	E	F	G	H	I	J	K	L	M	
4	7	0	0	6	1	3	3	0	0	0	0	0	
						(1)							
						+	2						Step 5
4	7	0	0	6	1	7	5	0	0	0	0	0	

Step 6: Multiply 7 on G by 7 on rod B and subtract the product 49 from rods HI. This leaves quotient 617 on EFC and remainder 1 on rod I. (Fig. 6)

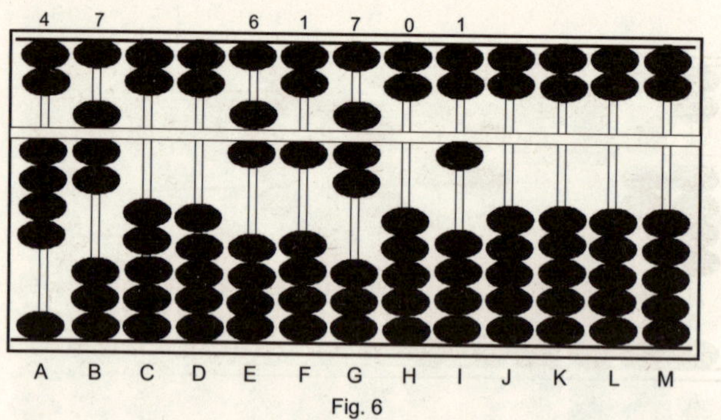

Fig. 6

Step 6

A	B	C	D	E	F	G	H	I	J	K	L	M	
4	7	0	0	6	1	7	5	0	0	0	0	0	
						−	4	9					Step 6
4	7	0	0	6	1	7	0	1	0	0	0	0	

Step 7: Next compare 4 on A and 1 on I. Follow rule: 1/4=2 plus 2 ⇒ The next number in the quotient will be 2. Place 2 on rod and add the remainder 2 on J. (Fig. 7)

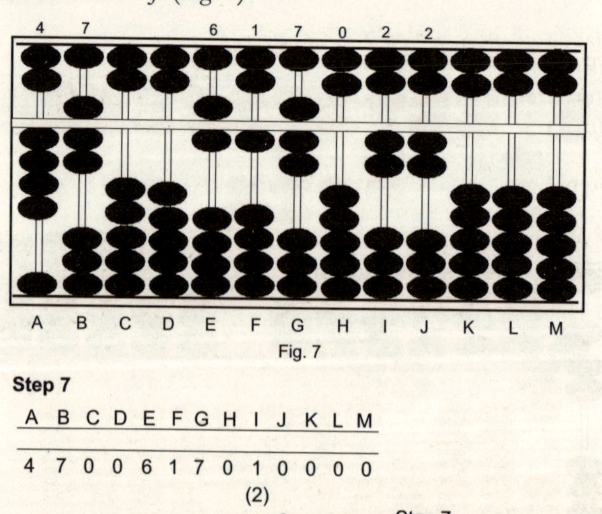

Fig. 7

Step 7

A	B	C	D	E	F	G	H	I	J	K	L	M	
4	7	0	0	6	1	7	0	1	0	0	0	0	
								(2)					
								+	2				Step 7
4	7	0	0	6	1	7	0	2	2	0	0	0	

Step 8: Multiply 2 on I by 7 on B and subtract the product 14 from rods JK. This is as far as we'll go in this problem. The answer so far is 6.1702 on EFGHI and the remainder 0.0006 on rod K. (Fig. 8)

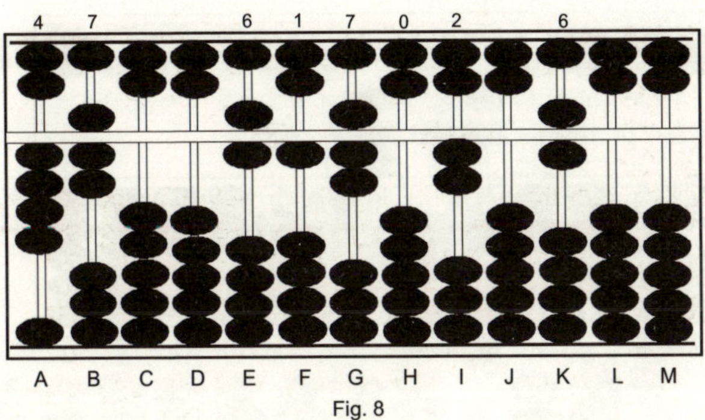

Fig. 8

Step 8

A	B	C	D	E	F	G	H	I	J	K	L	M	
4	7	0	0	6	1	7	0	2	2	0	0	0	
								−	1	4			Step 8
4	7	0	0	6	1	7	0	2	0	6	0	0	

Example 7: To show 28÷73=0.3835 ⇒ with remainder 0.0045 (see Fig. 1). This example uses the rules;

(a) rule 2/7=2 plus 6
(b) rule 6/7=8 plus 4
(c) rule 2/7=2 plus 6
(d) rule 4/7=5 plus 5

Step 1: Set the dividend 28 on rods EF and divisor 73 on AB Fig. 1).

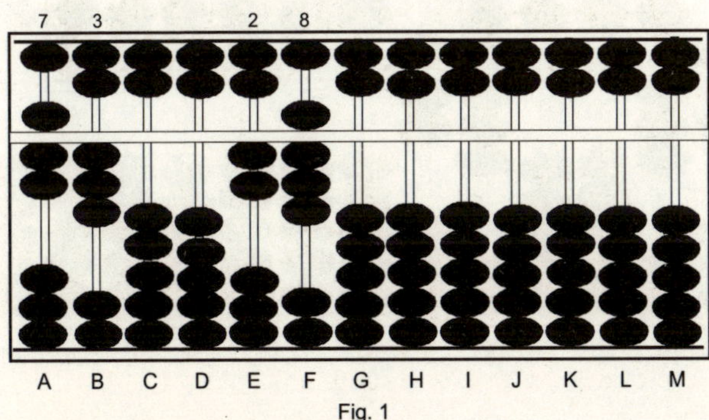

Fig. 1

Step 1

A	B	C	D	E	F	G	H	I	J	K	L	M	
7	3	0	0	2	8	0	0	0	0	0	0	0	Step 1

Step 2: Compare 7 on A and 2 on E. Follow rule: 2/7=2 plus 6 ⇒ The first number in our quotient answer will be 2. Place 2 on rod E and add the remainder 6 to rod F. (Fig. 2)

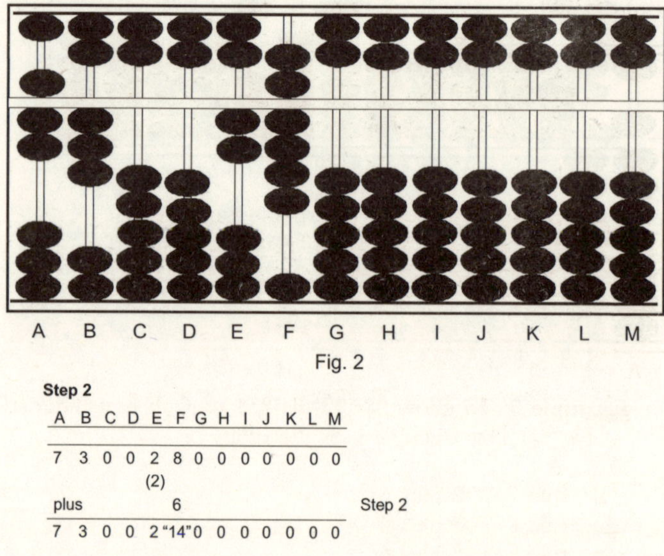

Fig. 2

Step 2

	A	B	C	D	E	F	G	H	I	J	K	L	M	
	7	3	0	0	2	8	0	0	0	0	0	0	0	
					(2)									
plus						6								Step 2
	7	3	0	0	2	"14"	0	0	0	0	0	0	0	

Step 3: With 14 on rod F we need to revise the answer. Add 1 to the quotient on rod E and subtract a further 7 from F leaving 37. (Fig. 3)

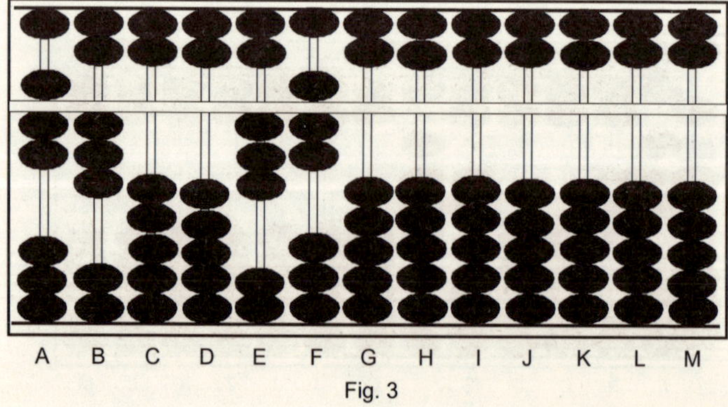

Fig. 3

Step 3

A	B	C	D	E	F	G	H	I	J	K	L	M	
7	3	0	0	2	"14"	0	0	0	0	0	0	0	
				+	1								
plus				–	7								Step
7	3	0	0	3	7	0	0	0	0	0	0	0	

Step 4: Now multiply 3 on rod E by 3 on B and subtract the product 09 from rods FG. This leaves the first number in our quotient 3 on rod E and the remainder 61 on FG. (Fig. 4)

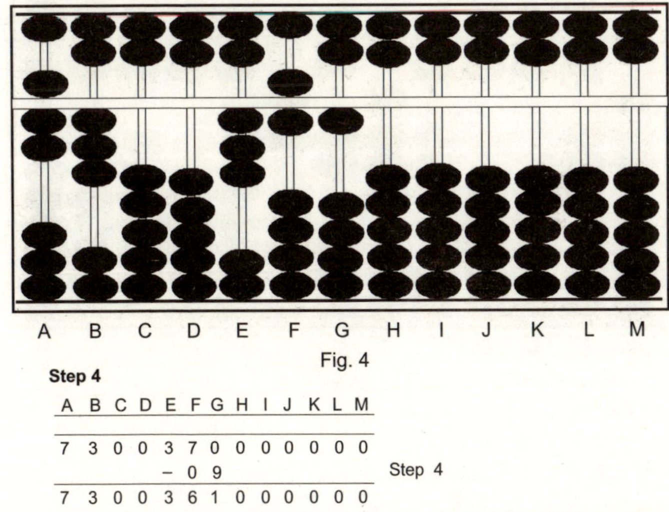

Fig. 4

Step 4

A	B	C	D	E	F	G	H	I	J	K	L	M	
7	3	0	0	3	7	0	0	0	0	0	0	0	
				–	0	9							Step 4
7	3	0	0	3	6	1	0	0	0	0	0	0	

Step 5: Now compare 7 on A with 6 on F. Follow rule: 6/7=8 plus 4 ⇒ The next number in our quotient will be 8. Place 8 on rod F and add the remainder 4 to G. (Fig. 5)

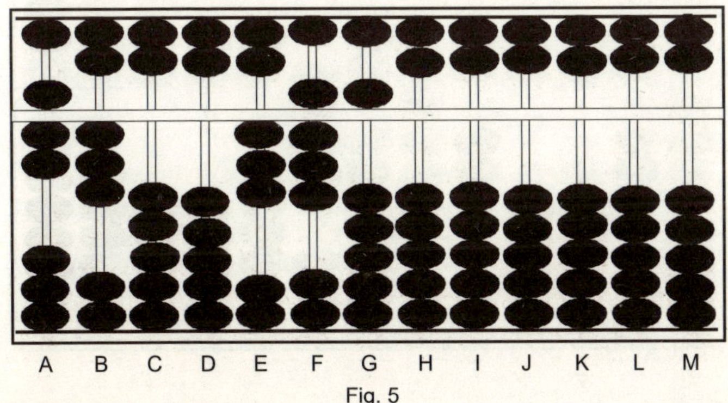

Fig. 5

Step 5

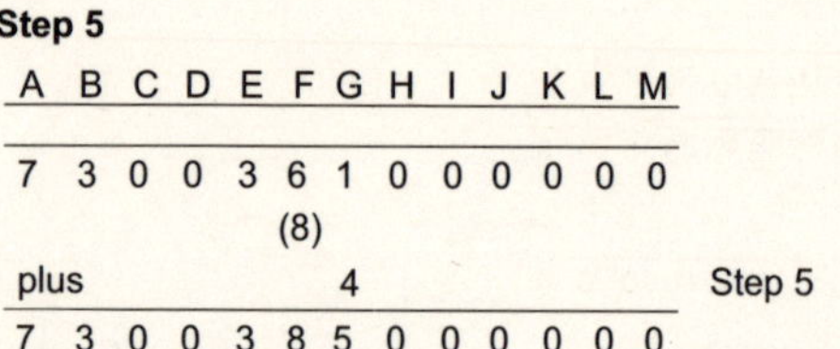

	A	B	C	D	E	F	G	H	I	J	K	L	M	
	7	3	0	0	3	6	1	0	0	0	0	0	0	
						(8)								
plus							4							Step 5
	7	3	0	0	3	8	5	0	0	0	0	0	0	

Step 6: Now multiply 8 on F by 3 on B. and subtract the product 24 from rods GH. This leaves the quitient 38 on EF and the remainder 26 on GH. (Fig. 6)

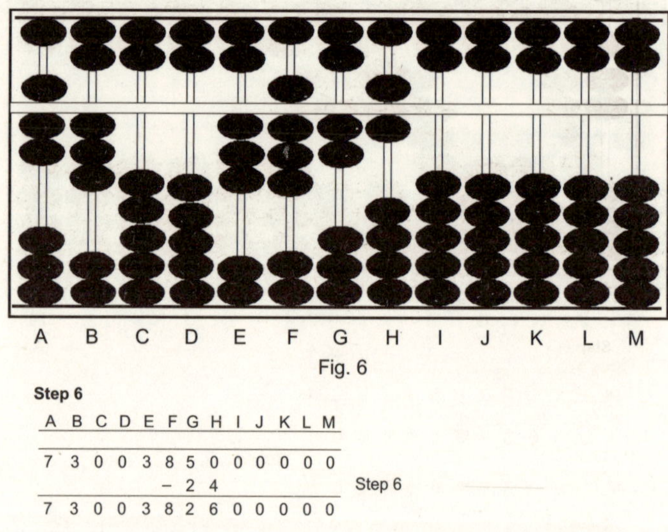

Fig. 6

Step 6

A	B	C	D	E	F	G	H	I	J	K	L	M	
7	3	0	0	3	8	5	0	0	0	0	0	0	
					−	2	4						Step 6
7	3	0	0	3	8	2	6	0	0	0	0	0	

Step 7: Compare 7 on A with 2 on G. Follow the rule: 2/7=2 pIus 6⇒ The next number in our quotient will be 2. Place 2 on G and add the remainder 6 to H. (Fig. 7)

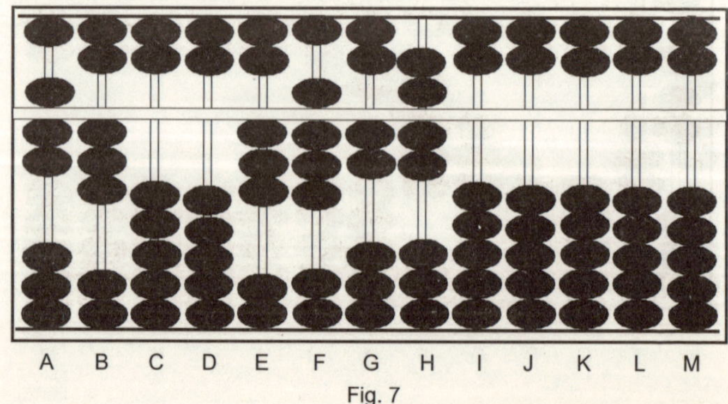

Fig. 7

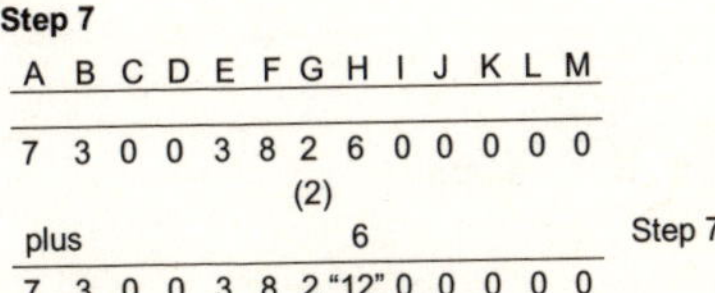

Step 7

A	B	C	D	E	F	G	H	I	J	K	L	M	
7	3	0	0	3	8	2	6	0	0	0	0	0	
						(2)							
plus							6						Step 7
7	3	0	0	3	8	2	"12"	0	0	0	0	0	

Step 8: With 12 on rod H we need to revise. Add 1 to the quotient on G and subtract a further 7 from H. (Fig. 8)

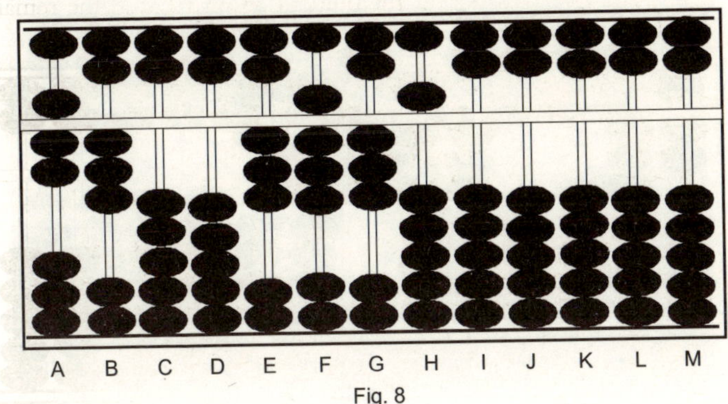

Fig. 8

Step 8

A	B	C	D	E	F	G	H	I	J	K	L	M	
7	3	0	0	3	8	2	"12"	0	0	0	0	0	
						+ 1							
							− 7						Step 8
7	3	0	0	3	8	3	5	0	0	0	0	0	

Step 9: Now multiply 3 on G with 3 on B and subtract 09 from rods HI. This leaves the quotient 383 on EFG and the remainder 41 on rods HI. (Fig. 9)

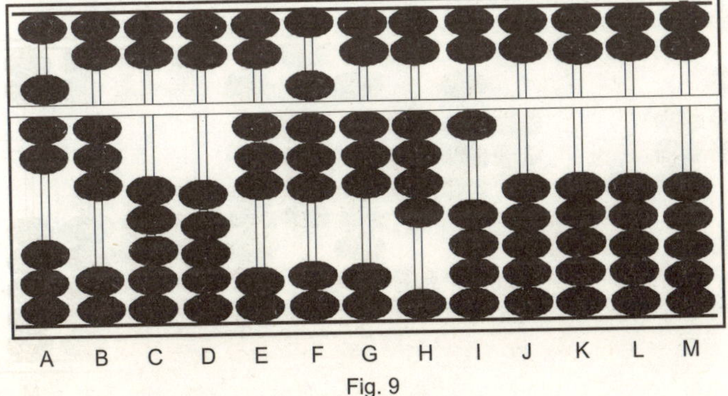

Fig. 9

Step 9

A	B	C	D	E	F	G	H	I	J	K	L	M	
7	3	0	0	3	8	3	5	0	0	0	0	0	
							− 0	9					Step 9
7	3	0	0	3	8	3	4	1	0	0	0	0	

Step 10: Compare 7 on A with 4 on H. Follow the rule: 4/7=5 plus 5 ⇒ The next number in our quotient will be 5. Place 5 on H and add the remainder 5 to rod I. (Fig. 10)

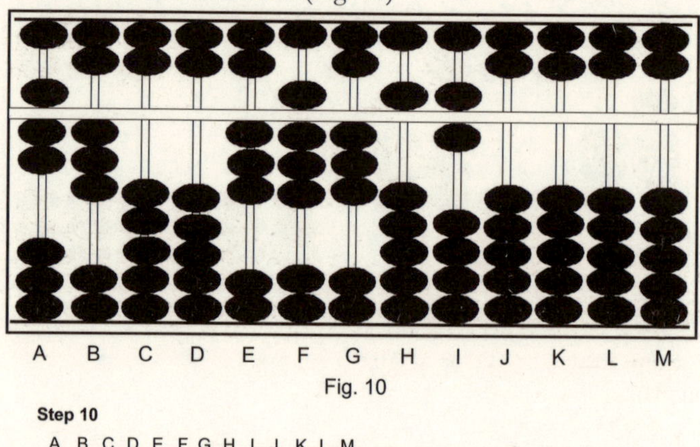

Fig. 10

Step 10

	A	B	C	D	E	F	G	H	I	J	K	L	M	
	7	3	0	0	3	8	3	4	1	0	0	0	0	
								(5)						
plus									5					Step 10
	7	3	0	0	3	8	3	5	6	0	0	0	0	

Step 11: Now multiply 5 on H with 3 on B and subtract 15 from rods IJ. This is as far as we'll go in this problem. The answer so far is 0.3835 on EFGH and the remainder 0.0045 on rods IJ. (Fig. 11)

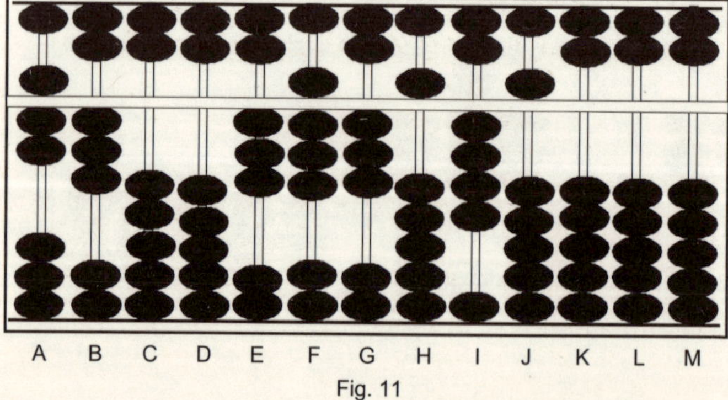

Fig. 11

Step 11

A	B	C	D	E	F	G	H	I	J	K	L	M	
7	3	0	0	3	8	3	5	6	0	0	0	0	
							−	1	5				Step 11
7	3	0	0	3	8	3	5	4	5	0	0	0	

Thus, 28÷73=0.383545.

Note: The following examples uses the rules:

(a) rule 5/9=5 plus 5
(b) rule 8/9=8 plus 8
(c) rule 9/9=forward 1

Example 8: Show that 58174÷986=59 on an abacus.
This example uses the rules:

(a) rule 5/9=5 plus 5
(b) rule 8/9=8 plus 8
(c) rule 9/9= forward 1

Step 1: Set the dividend 58174 on rods FGHIJ and the divisor 986 on ABC. (Fig. 1)

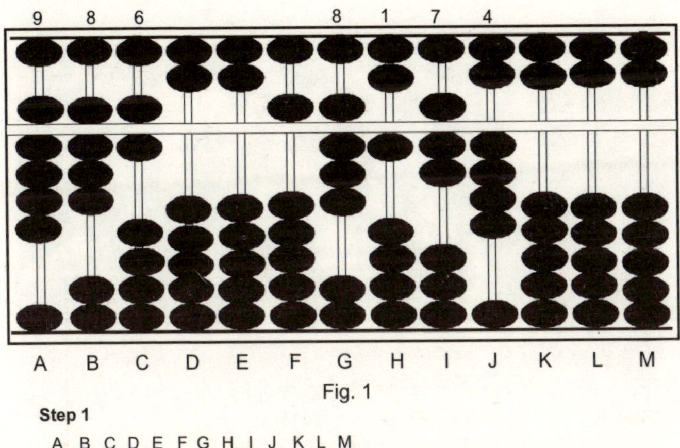

Fig. 1

Step 1

A	B	C	D	E	F	G	H	I	J	K	L	M	
9	8	6	0	0	5	8	1	7	4	0	0	0	Step 1

Step 2: Follow rule 5/9=5 plus 5 ⇒ place the quotient 5 on rod F and the remainder 5 on G, leaving 5 "13" 174 on FGHIJ.

(*Note*: This is where we use the *The Extra Bead Technique* which allows for 13 to be placed on rod G.) (Fig. 2)

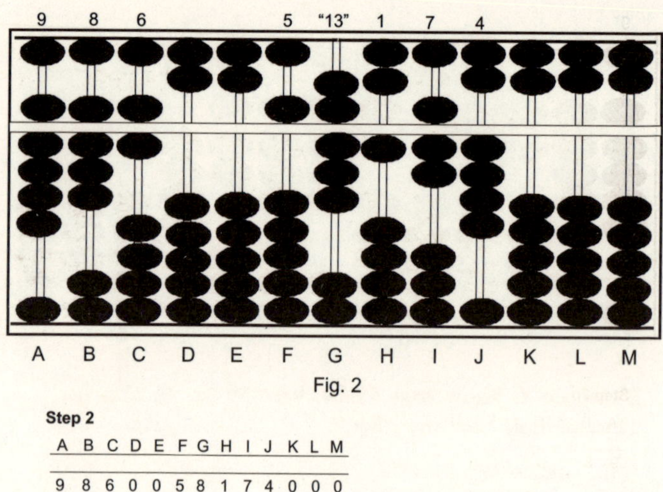

Fig. 2

Step 2

A	B	C	D	E	F	G	H	I	J	K	L	M	
9	8	6	0	0	5	8	1	7	4	0	0	0	
plus						5							Step 2
9	8	6	0	0	5	"13"	1	7	4	0	0	0	

Step 2a: Multiply 5 on F by 8 on B. Subtract the product 40 from rods GH, leaving, 59174 on rods FGHIJ. (Fig. 3)

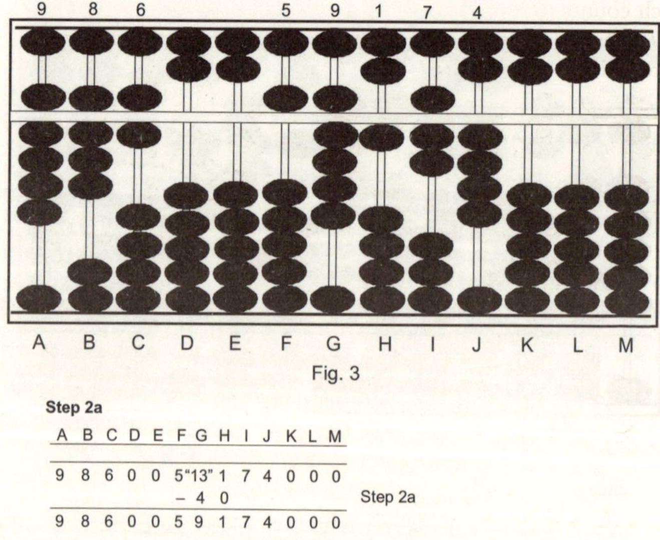

Fig. 3

Step 2a

A	B	C	D	E	F	G	H	I	J	K	L	M	
9	8	6	0	0	5	"13"	1	7	4	0	0	0	
						– 4	0						Step 2a
9	8	6	0	0	5	9	1	7	4	0	0	0	

Step 2b: Multiply 5 on F by 6 on C. Subtract the product 30 from rods HI, leaving, 58874 on rods FGHIJ. (Fig. 4)

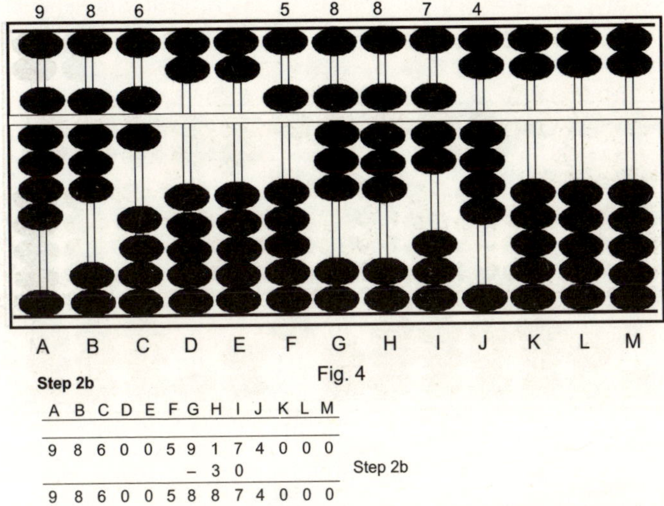

Fig. 4

Step 2b

A	B	C	D	E	F	G	H	I	J	K	L	M	
9	8	6	0	0	5	9	1	7	4	0	0	0	
						–	3	0					Step 2b
9	8	6	0	0	5	8	8	7	4	0	0	0	

Step 3: Follow rule 8/9=8 plus 8 ⇒ place the quotient 8 on G and add the remainder 8 on H.

(Note: Since rod H already carries a large number we don't have 8 available. This is where we'll use *The Suspended Bead Technique* which counts for 10.) *Slide the top most bead ½ way down on rod H for a value of 10* then subtract 2 giving us the required 8. This leaves 5 8 "16" 7 4 on rods FCHIJ. (Fig. 5)

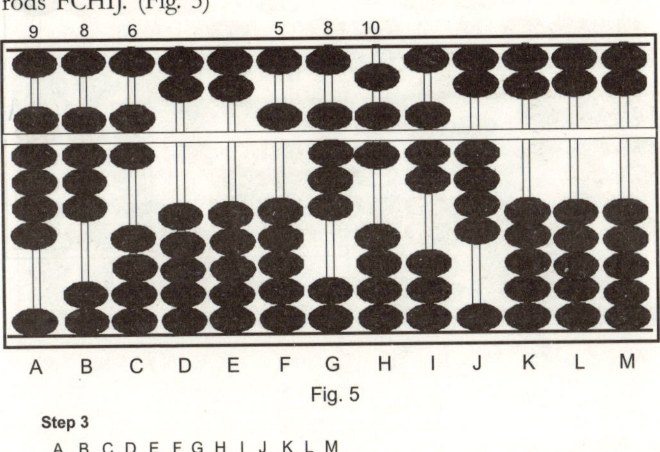

Fig. 5

Step 3

A	B	C	D	E	F	G	H	I	J	K	L	M	
9	8	6	0	0	5	8	8	7	4	0	0	0	
suspended							"10"						
						–	2						Step 3
9	8	6	0	0	5	8	"16"	7	4	0	0	0	

Step 3a: Multiply the 8 on G by the 8 on B and subtract the product 64 from rods HI, leaving 5 8 "10" 3 4 on FCHIJ. (Fig. 6) ***see a variation below***

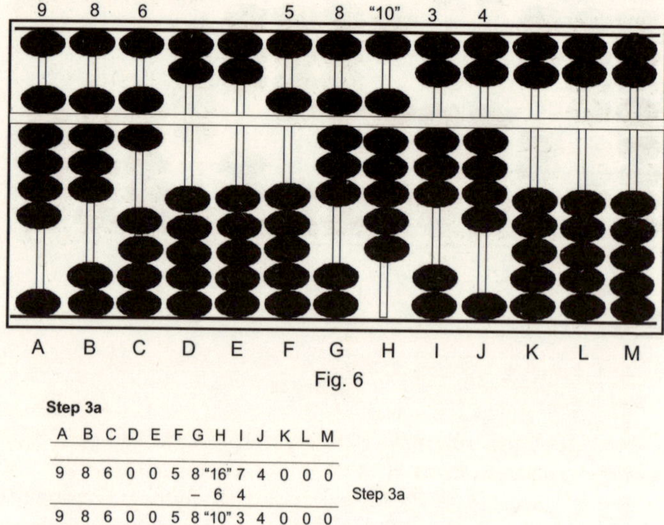

Fig. 6

Step 3a

A	B	C	D	E	F	G	H	I	J	K	L	M	
9	8	6	0	0	5	8	"16"	7	4	0	0	0	
						−	6	4					Step 3a
9	8	6	0	0	5	8	"10"	3	4	0	0	0	

Step 3b: Multiply the 8 on G by the 6 on G and subtract the product 48 from rods IJ, leaving 58986 on rods FGHIJ. (Fig. 7)

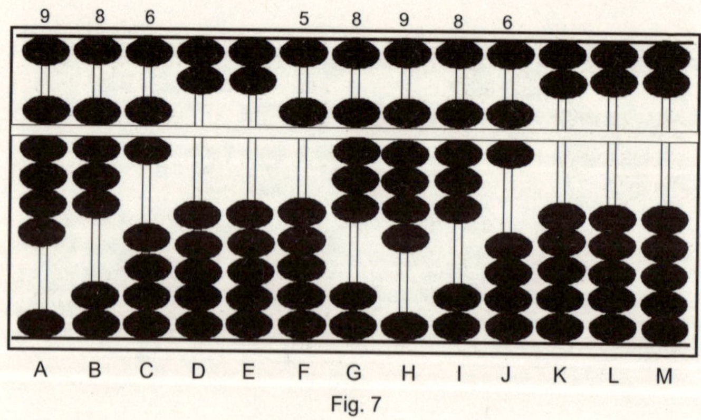

Fig. 7

Step 3b

A	B	C	D	E	F	G	H	I	J	K	L	M	
9	8	6	0	0	5	8	"10"	3	4	0	0	0	
						−	4	8					Step 3b
9	8	6	0	0	5	8	9	8	6	0	0	0	

Step 3c: Here we make a revision. Changing the 8 on G to a 9 we can now subtract a further 986 from rods HIJ. This leaves 59 on rods FG which is the answer. (Fig. 8)

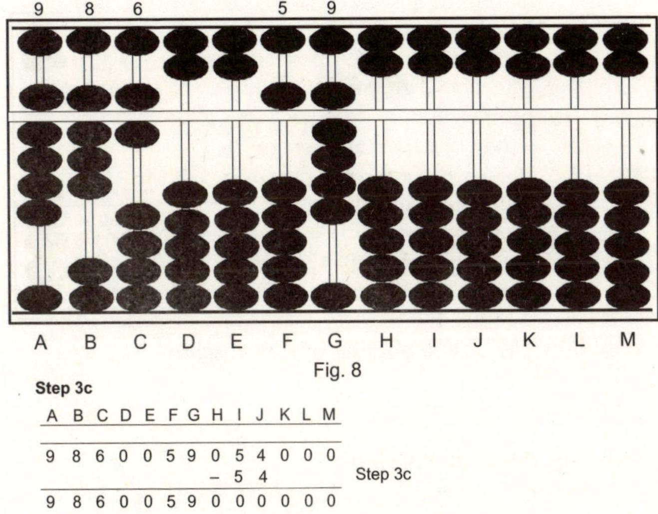

Fig. 8

Step 3c

A	B	C	D	E	F	G	H	I	J	K	L	M	
9	8	6	0	0	5	9	0	5	4	0	0	0	
							−	5	4				Step 3c
9	8	6	0	0	5	9	0	0	0	0	0	0	

Note: In the above step 3a one could have exercised the option to do a revision using rule 9/9=forward 1⇒ thereby changing the 8 on rod G to a 9. Multiplying the revised 9 on G by the 8 on B followed by the 6 on G and subtracting the products from rods HIJ, one could have come to the answer in a slightly different way.

Example 9: Find on abacus, 3804/12=?

This Example is described below without diagrams. Try it on abacus. The bead movement is described below.

	3804	
3	–3	The first digit of the divisor is 1, 1×3=3, take 3 from first digit
	0804	is the first partial dividend
	–6	The second digit of the divisor is 2, 1×3=6 take six from second digit
	204	is the second partial dividend
32	–2	The divisors's first digit yield a factor of 2 (the multiple take 2
	004	is the third partial dividend
32*****	–4***	The divisor's second digit is 2, multiply by the multiple 2 to yield 4. Problem!! Can't take 4 off so backtrack

–1	104	Remove 1 from the multiple, and multiply it by the flirt divisor digit. So we arrive at this revised partial dividend
31	–2	Therefore most recent digit of the multiple, 1, is multiplied by the second digit of the divisor 2, and taken away
	84	is our next partial dividend.
313	–8	The first digit of the divisor, 1, has a multiple of cc subtract 8
	04	is our next partial dividend.
318*****	–16**	The divisor's second digit is 2, so 8×2=16, subtract 16 Problem again! Backtrack
–1	14	And recover 1×1 (first divisor digit * recovered multiple)
317	14	It is our revised partial dividend
	–14	The second digit of the divisor is 2, therefore, 7×2=14, subtract 14
	00	This leaves nothing on the abacus.

The answer is therefore **317**.

Manipulating decimal division

We can force a decimal divisor into an integral on just by scaling by moving the decimal point. However, the number and direction of places the decimal point moves achieve this must be matched also by the other number to balance out the effect of this change.

E.g.
462.1/3.9=(4621/39)=(46210/390)=(46.21/0.39)=(4.621/0.039)

And so on. We note that each differs by factors of 10 from each other, because of the way we have move the decimal point. Once one numl.r has been changed, the other must follow in the same manner. This keeps the ratios constant and allows us to manipulate the numbers before we load them onto the abacus.

To some, this is not very satisfactory, because we may have two numbers with virtually endless number of decimal significant places. To ascertain the correct location of the decimal requires one to think about the number of digits and their decimal point positions.

11/11000 = 0.001
11/1100 = 0.010
11/110 = 0.100
11/11 = 1.000
11/1.1 = 10.000
11/0.11 = 100.000

11/0.011 = 1000.000
220/11000= 0.02
220/1100 = 0.20
220/110 = 2.00
220/11 = 20.00
220/1.1 = 200.00
220/0.11 = 2000.00
220/0.011 = 20000.00

Square Root

Abacus can be used to solve more difficult computational problems such as square root and cubic root. The procedures used in 7 solving these problems again remind us of algorithms for electronic computers: simple procedures are repeated many times to solve complex problems.

Square Root Instructions

For simplicity, we give instructions for integers only. It is easy to generalize to any positive real number.

1. Add the integer number on the abacus. We shall call this number the operand.
2. Group the digits into columns of two for right to left. For example, 509 should be grouped as 5 09. And, 94664 should be grouped as 9 46 64.
3. Start from the first group from left, add 1 to the result, and subtract one from the first group.
4. If the first group is greater than result plus 2, add 2 to the result and subtract the value of the result from the first group. Continue adding 2 to result and subtracting from the first group until the value for the first group is smaller than the result.
5. If there are more groups, shift one column to the right for the result and 1 group (2 columns) right for the operand.
6. Add 11 to the result and subtract the result from the operand. If after adding 11 the result is greater than the operand, then don't add 11 but continue shifting for the result and the operand. Then add 101 or 1001 etc. to the result and subtract the result from the operand.
7. Add 2 to the result and subtract the result from the operand. Continue adding 2 to result and subtracting from the operand until the value is smaller than the result.
8. Repeat steps 5-7 if necessary.
9. The final square root value is (result+1)/2.

Example 1: $\sqrt{529}=23$

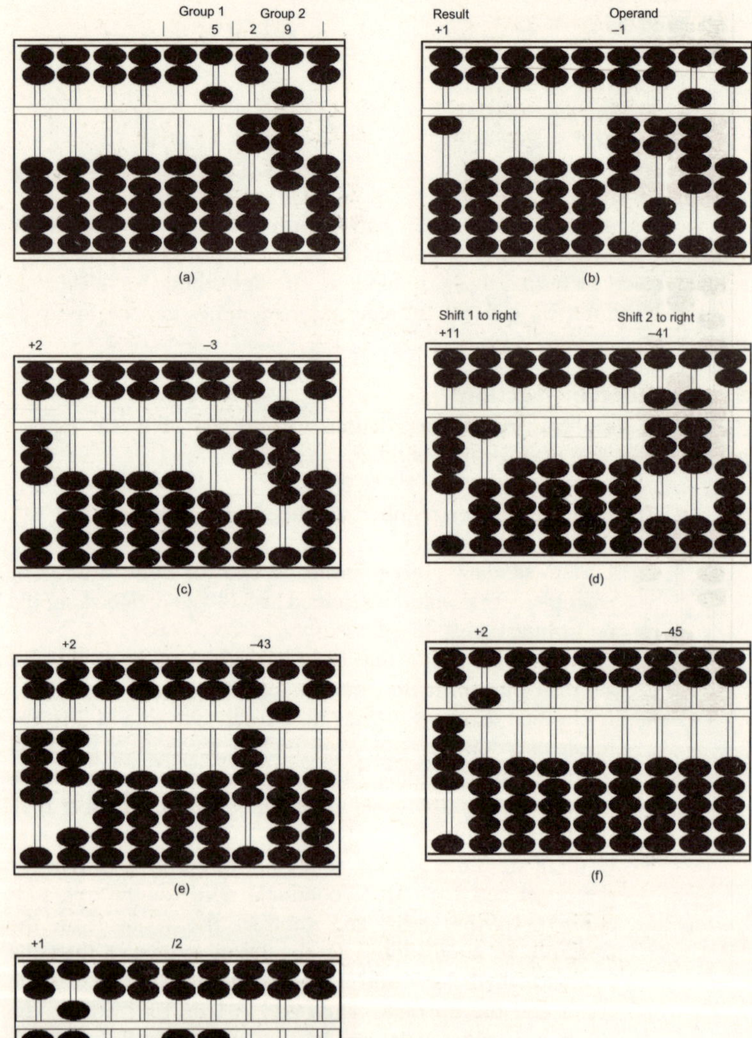

(a) (b) (c) (d) (e) (f) (g)

Example 2: $\sqrt{94864}=308$

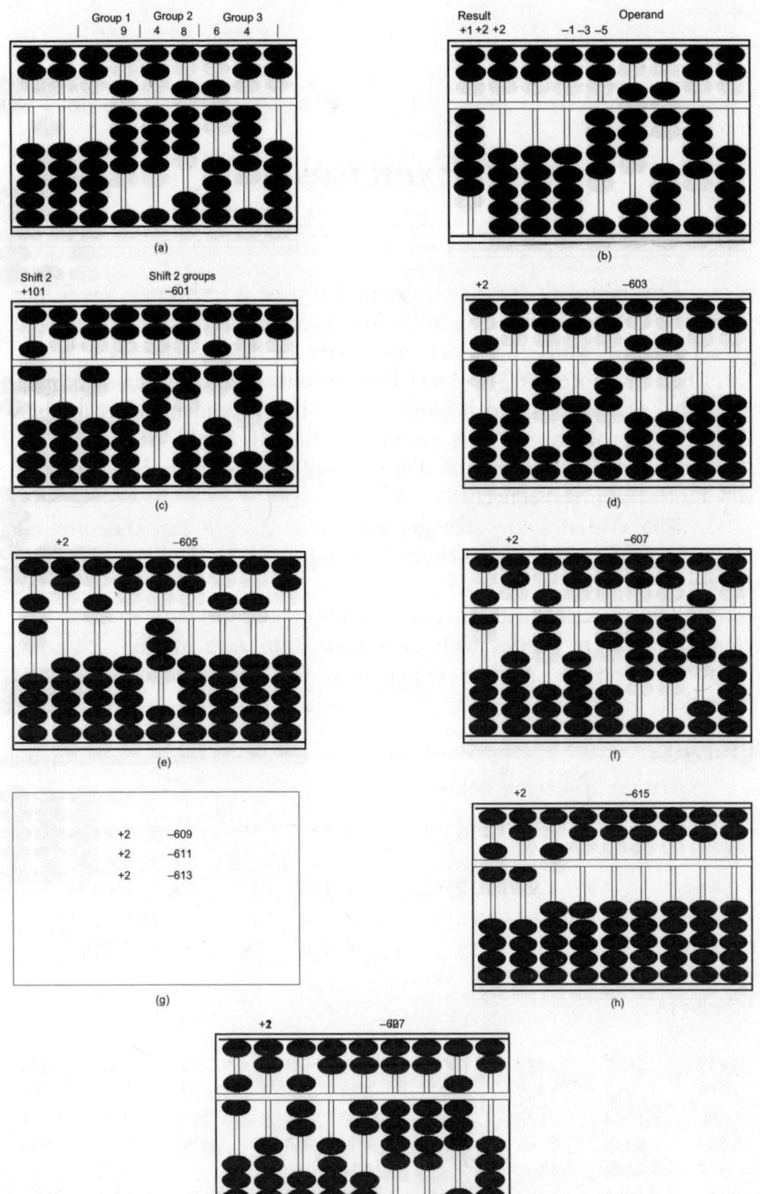

(a) (b) (c) (d) (e) (f) (g) (h) (i)

Exercises

Constant daily practice is essential if- one is to become proficient in the use of the abacus. The following exercises, prepared and arranged in accordance with ,the most up-to-date methods, have ben kindly furnished by Professor MiyokicH Ban, an outstanding abacus authority. They will provide a good beginiing forh serious student, wh cap then find more probtems in any ordinary rfthrneti book. AL.sojiete th4 pr6blems in muItiplication and division may be used as probifflis i ditior an4 Libtraction. respectively.

The exercises are arrangei so that a student can measure his progress against the yardstick of the Japanese or Chinese licensing system.

The exercises (with answers) are also chosen to ietbe nximiwt1 qf variety 4o the' problems, with each digit from zero to rine. receiving equal attention an essential requirement for improvement in abacus operation.

Exercises

Try the following below:

- 1+2=3
- 2+3=5
- 3+4=7
- 5+7=12
- 6+6=12
- 8+9=17
- 7+8=15
- 9+7=16
- 7+5=12
- 8+13=21
- 50+85=135
- 6+7=13
- 25+13=38
- 4+27=31
- 23+20=43
- 2+8=10
- 6+9=15
- 31+7=38
- 3+8=11
- 9+5=14
- 21+8=29
- 76+82=158
- 8+6=14

1339	4649	7420	1005	3080	4196	2384	9521	2039
1415	6917	1459	742	898	8978	3129	3128	9169
686	4743	2128	1280	4748	9855	343	4798	5611
6483	6035	7944	8361	3408	6776	7048	1137	+700
+978	+3368	+4599	+7513	+1826	+790	+9572	+9752	+731
10901	25712	23550	18901	22051	30595	22476	28336	18250

6815	6920	9113	7288	2514	9652	7431	8894	1495	8931
7128	9653	2704	300	3563	8985	8519	8985	7130	7644
8733	5043	6692	9819	2458	4093	3941	2744	6779	7084
9576	7847	1123	2449	8162	1543	4572	8552	1400	2405
+7946	+5073	+7864	+5991	+4619	+3580	+3749	+3208	+6342	+3626
40198	34536	27496	25847	21316	27863	28212	32380	23146	29490
9253	4288	3547	5200	8250	4954	9260	8262	7290	3573
1375	2872	5879	4044	2809	5000	5064	4600	1072	4517
9572	2745	3915	6730	2565	8858	4104	8855	1947	5267
8768	5458	7314	8967	345	500	4069	6021	9626	5406
+5848	+1243	+6106	+5311	+3631	+669	+3567	+8711	+5911	+3969
34816	16105	26761	30252	17600	19981	26064	36449	25846	22732
7143	2866	705	3606	1573	5743	2468	1948	8835	2065
5649	4514	9981	3750	204	7483	8577	2120	4804	8662
7739	2038	1141	5574	5027	734	8163	9596	9280	4195
1566	6814	9733	710	4164	6108	9682	1593	5706	750
+4307	+7779	+4744	+8512	++4401	+7512	+5218	+9833	+9148	+9326
26404	24011	26304	22152	15372	27580	34108	25090	37773	24998
4929	5832	5327	5101	7563	7452	3350	2677	8322	664
5208	9288	7194	3163	2148	2167	3370	7972	604	1288
4285	1581	7487	3052	5752	2258	2024	361	349	5300
4213	9535	9029	1351	6649	5163	4693	2028	147	9262
+9659	+9342	+1178	+8030	+4572	+9304	+2035	+576	+2485	+6758
28294	35578	30215	20697	26684	26344	15472	13613	12206	23602
6413	2518	2473	1961	2635	7881	8454	9373	4254	7989
7423	4180	2576	4282	3296	7912	3498	9068	4110	4430
9611	1427	1627	4355	8276	2916	6076	2445	5541	8184
922	8831	6371	8415	8429	9832	7059	5185	9217	9977
+3778	+7781	+3459	+2610	+9188	+7044	+489	+7277	+9728	+9231
27247	24737	16506	21623	31824	35585	25576	33351	32850	39811
4268	2403	7347	7558	5621	5214	8764	5406	7177	8939
2839	1528	5406	3738	2033	5356	8557	814	2481	8440
1171	5889	139	9525	2061	4409	7513	2540	3965	872
5121	5474	1998	5490	9842	8378	8185	8647	1627	3622
+4725	+6046	+9385	+5461	+8884	+5299	+3832	+7609	+6497	+9282
18124	21340	24275	31772	28439	28656	36853	25013	21747	31155

5913	2748	5501	7464	574	8847	7413	8406	5305	2995
1482	9426	1481	6528	2092	7365	5388	3872	6739	4893
3256	6022	638	3384	8251	3091	3767	8881	8842	2825
3105	8156	5646	4220	7047	1701	2060	4677	4371	9801
+285	+9470	+9913	+4575	+3675	+3329	+6316	+3087	+9288	+5553
14041	35819	23179	26171	21639	24333	25364	28923	34545	26067

8582	8686	3573	5857	1013	6088	836	9329	3295	7699
4225	6226	7963	1709	638	4993	9897	7999	9807	8068
724	1266	8989	6407	6239	266	4205	4796	5598	2841
7719	4632	5825	9940	5890	5936	4499	2095	9589	4504
+7079	+7858	+8011	+3355	+4633	+4333	+7377	+9668	+5780	+1715
28329	28668	34361	27268	17413	21616	26714	33887	34069	24827

357	420	6980	1380	5092	8567	5499	829	430	4761
3856	9857	2299	4130	1981	7823	1707	5498	3434	4755
7287	4793	6509	1495	3222	4870	5962	9873	8670	2337
9208	9314	1594	2029	5302	2377	5511	4408	1535	4893
+3854	+8090	+3193	+4673	+3126	+2347	+8846	+8350	+4327	4311
24562	32474	20575	13707	18723	25984	27525	28958	18939	21057

9654	2019	6411	9534	176	3670	9662	6612	737	3279
6996	3313	839	6086	5222	5210	7124	6295	214	1291
8742	8116	9952	9056	274	7080	2536	9976	5505	2093
9194	177	8765	120	9686	4634	5825	9477	3243	3697
+6584	+560	9816	+3610	+6080	+3434	+9986	+204	+4099	+2566
41170	14183	35783	28406	21438	24028	32564	13798	12926	

5129	9408	4112	339	6590	3783	5057	6104	1973	4125
8069	3515	9153	6173	4399	2933	2583	9421	9805	1763
4531	1875	4728	875	7708	1546	9791	1602	3739	3385
6572	9761	3582	9915	9336	7784	6484	9740	911	9867
+1912	+9763	+6085	+7233	+5956	+7160	+1935	+2862	+2371	+1004
26213	34322	27660	24535	33989	23206	25854	29729	18799	20144

35	14	57	62	20	49
70	20	69	15	38	80
29	97	32	40	92	21
86	65	70	71	13	68
43	28	31	84	56	32
58	73	46	26	90	14
10	41	18	97	84	95
64	56	94	80	47	70
17	30	25	39	75	53
92	89	80	53	61	76
504	513	522	567	576	558

360	106	528	967
829	543	160	239
213	928	427	650
308	710	951	108
497	954	719	243
932	329	452	758
589	267	106	491
690	695	843	536
147	514	385	702
674	870	690	871
850	308	724	460
201	796	381	629
765	632	201	984
148	807	579	315
756	481	634	870
7959	8940	7782	8832

Subtraction

Exercises

Try the following out:

• 86–72=14	• 4–2=2	•5–3=2	•6–2=4	•8–6=2
• 32–19=13	•17–4=13	•11–6=5	•15–6=9	•12–6=6
• 24–17=7	• 88–5=43	•40–15=25	•30–22=8	•27–12=15
• 102–39=63	•136–74=62	•112–50=62	•300–126=174	•10000–41=959
• 528–89=439				
• 941–812=129				
• 160–74=86				

49576	30781	49507	60964	51259	61777	45804	53626	41514	43337
–1537	–1053	–2441	–6449	–967	–3144	–416	–3246	–8185	–1898
–4400	–650	–6926	–9942	–2847	–9558	–6872	–9543	–3575	–4947
–3700	–2594	–4104	–6002	–9039	–8964	–3987	–2284	–1435	–4249
–6423	–8512	–6942	–7890	–5333	–7442	–5891	–1193	–5004	–2366
–2197	–5537	–4560	–7954	–8680	–4534	–3962	–5253	–8577	–6692
–7501	–4615	–1911	–8094	–6066	–5872	–2950	–8943	–6193	–5118
–6881	–990	–8258	–3513	–853	–9080	–6368	–5704	–2246	–8529
–7436	–1868	–1117	–4979	–7311	–5204	–7012	–8680	–218	–2326
–8341	–1192	–6830	–5717	–8157	–3513	–3390	–7790	–1948	–553
1160	3770	6418	424	2006	4466	4956	990	4135	6659

52479	51563	42082	49760	55487	48228	49906	46311	65081	42632
–7668	–7909	–6931	–5489	–4153	–7743	–4299	–2818	–9711	–961
–3968	–5129	–6810	–6224	–2839	–3856	–6886	–6261	–3901	–3531

–3084	–2767	–6888	–7149	–5974	–5964	–2849	–2602	–9746	–3992
–2556	–3395	–1102	–2127	–3634	–6789	–8141	–2962	–9409	–5633
–1158	–8841	–534	–4265	–9623	–1507	–5959	–7106	–2335	–827
–6031	–9800	–3358	–3486	–8350	–4650	–3397	–7612	–4223	–7338
–4580	–853	–4302	–7132	–2937	–838	–1541	–967	–6642	–7975
–8026	–1010	–5637	–4248	–4745	–5038	–7040	–8347	–9985	–385
–8847	–7404	–4478	–697	–5805	–6989	–4225	–1922	–3003	–9286
6561	4455	2042	8943	7427	4854	5569	5714	6126	2703

44585	49183	46071	48486	52302	44753	59480	37147	46346	52957
–2219	–6215	–4776	–1968	–2821	–4092	–9457	–3275	–7337	–4723
–7966	–7135	–8515	–1018	–8168	–1811	–9759	–318	–9596	–4248
–7794	–177	–651	–9960	–5799	–5296	–8119	–1391	–730	–3406
–9598	–5908	–3998	–9136	–4173	–3773	–6612	–2436	–8945	–8946
–477	–2480	–4177	–636	–3716	–3238	–3603	–754	–927	–6792
–2080	–7775	–5159	–3842	–7022	–1783	–619	–9528	–4421	–6784
–9161	–1246	–1741	–8773	–4679	–9364	–8219	–3306	–4169	–6989
–1218	–7159	–9528	–5673	–9215	–6764	–5634	–9624	–5769	–3075
–2596	–6644	–1282	–3557	–2592	–6963	–3360	–4172	–313	–3870
1476	4444	5944	3923	4117	1669	4098	2343	4139	4124

46245	46989	45382	55450	59895	58246	46617	59770	54619	51186
–1076	–6957	–2774	–5777	–9175	–7812	–8658	–7833	–9014	–6969
–2259	–840	–591	–3252	–1531	–6517	–3068	–5895	–5825	–5871
–2806	–6196	–5320	–5410	–1717	–6779	–5293	–6389	–1560	–3601
–666	–7521	–3411	–6972	–273	–5612	–8607	–9966	–8679	–8239
–9811	–4002	–5715	–116	–7944	–8567	–203	–5068	–4941	–1130
–6711	–5441	–324	–7217	–9302	–6604	–3437	–8720	–251	–6930
–304	–5381	–9524	–8233	–4868	–1415	–4601	–22848	–3623	–1841
–8898	–6544	–8709	–1460	–9932	–5602	–5768	–592	–7382	–2419
–4450	–1781	–5123	–9758	–7482	–823	–1501	–4554	–4631	–8410
9264	2326	3891	7255	7671	8515	5481	8505	8713	5776

45415	48960	31943	55510	68706	50940	68126	49128	54832	63774
–1756	–5061	–561	–6049	–8887	–4649	–9377	–2918	–5696	–3998
–8839	–5886	–8894	–8201	–3681	–8290	–6368	–7359	–9585	–6255
–6860	–2468	–631	–6916	–8998	–2632	–7444	–3202	–1279	–5888
–603	–208	–2216	–6156	–9062	–2462	–9534	–2059	–4819	–1815
–4061	–8974	–862	–9335	–8448	–2287	–373	–5091	–4350	–953
–4697	–3203	–4920	–132	–4195	–6208	–5861	–930	–8041	–8189
–2318	–8732	–6796	–3433	–5169	–740	–3648	–3203	–4521	–9834
–3210	–9032	–3674	–938	–1447	–9038	–9655	–6081	–8437	–9528
–3352	1654	–1418	–7712	–9289	–7900	–6221	–9723	–2235	–9205
9719	3742	2241	6638	9530	6734	9645	8562	5869	8111

51733	58482	57720	52327	52776	50433	47096	48016	50782	39898
–6178	–5546	–8271	–3563	–9852	–8263	–1706	–1059	–7739	–8121
–524	–9932	–3514	–5836	–203	–6866	–6325	–2517	–9279	–4611
–2489	–8624	–7032	–3225	–9697	–2235	–6598	–5374	–734	–3476
–8903	–7096	–7433	–3735	–8778	–5802	–8237	–9962	–4798	–7104
–3572	–4852	–7983	–7084	–8867	–1690	–3686	–8654	–2989	–1398
–4510	–4631	–6511	–4149	–1032	–6828	–7822	–4282	–4145	–5406
–3996	–1722	–7375	–6699	–4188	–6614	–6597	–2465	–9404	–3932
–6553	–7020	–2293	517	–7028	–2078	–5304	–194	–2663	–2669
–2791	–1469	–1399	–9322	–2875	–7891	–1597	–4182	–5058	–2559

7496	7590	5909	9147	256	170	1220	9327	3973	622
62930	50188	50895	49291	56887	60968	41270	49136	38100	47372
–4872	–7891	–9800	–608	–8986	–5081	–1750	–9119	–4207	–6056
–7232	–8483	–209	–7925	–6691	–5091	–771	–6694	–5338	–5112
–9049	–1285	–2130	–3579	–6689	–8249	–5560	–3968	–131	–5930
–2754	–503	–9321	–1790	–2439	–8748	–8004	–3517	–2425	–2445
–2499	–6274	–8588	–2508	–2471	–5884	–4366	–2298	–1244	–1111
–9374	–9245	–5631	–5539	–4086	–8574	–9450	–2821	–978	–7515
–3912	–462	–9364	–6354	–7604	–656	–5573	–2753	–2005	–9794
–7230	–4086	–2043	–8952	–6998	–5677	–1632	–5668	–7991	–6994
8799	8517	300	5467	5671	5864	1617	6517	7207	1503

44656	50901	55468	51468	66468	47068	65020	57911	39529	54766
–422	–9287	–8462	–693	–3587	–9290	–5667	–9296	–782	–7445
–4950	–7188	–229	–9673	–7361	–4922	–6506	–1439	–5199	–3516
–4456	–1803	–5072	–3478	–6382	–4136	–9473	–5294	–8643	–484
–8456	–3526	–6506	–326	–6853	–5008	–5392	–6790	–2762	–7845
–5681	–8982	–8319	–7689	–8929	–981	–3830	–6008	–6310	–3932
–6653	–3404	–2655	–2441	–6922	–8023	–8466	–8353	–3111	–5394
–3069	–3917	–7168	–4105	8462	–2744	–8923	–5504	–1119	–8100
–2062	–2180	–4279	–4432	–8419	–9276	–2566	–2920	–5323	–7869
–4985	–4089	–6521	–9281	–2745	–560	–8038	–4211	–5554	–4506
3922	6525	6257	9350	6808	2128	6159	8096	726	5695

45661	58292	58280	60738	41289	66315	52764	54524	54469	60163
–2644	–9067	–9849	–5005	–5025	–8676	–5026	–2332	–5449	–1794
–4929	–5013	137	–472	–7520	–1540	–4219	–5721	–6452	–7045
–7958	–9161	–711	–3740	–2497	–2465	–9531	–6302	–9722	–6393
–1986	–5952	–1166	–9805	–7122	–4085	–3716	–4358	–5569	–5330
–5562	–8870	–8040	–7175	–3133	–8975	–6622	–1917	–924	–2319
–9055	–6129	–4957	–8430	–3711	–8379	–665	–4616	–206	–9810
–2085	–1419	–7983	–4990	–283	8581	–8561	–7724	–9081	–3456
–8292	5134	–5738	–8200	–8450	–6566	–7485	–9671	–2674	–5891
–1821	–3649	–9955	–3236	–3070	–9605	–5954	–3795	–5921	–9574
1329	3896	9744	9685	475	7443	986	8088	8471	8551

Addition and Subtraction

5819	6489	5323	6135	4584	1207	3621	1026	1217	6625
9323	1235	6620	3743	8469	8390	4904	1755	4256	–2507
–1135	3857	5392	954	3446	–2321	8615	8921	8691	8285
2611	7227	–4245	5220	2899	5014	–161	–2059	2451	9868
7077	7156	2305	8552	–9189	4170	–9237	7559	8094	9868
23695	25964	11755	15139	12530	14345	12917	19100	20199	24722

4350	3428	8008	1073	6871	9785	8506	7977	1329	9381
–4067	4349	6361	7879	–2384	–6898	–5637	6331	490	–5352
9310	–6763	–5603	2177	8004	6492	–449	–6211	9353	7270
–6168	9590	2920	–4306	8018	–4334	3496	8622	6725	–7437
997	–7165	3929	8447	8186	1374	3354	–8379	5132	727
4422	3434	15615	15270	28695	6419	9270	8340	23029	4589

90	82	86
26	54	92
–83	–73	50
59	–38	29
42	10	–74
–15	29	51
–60	76	63
47	–61	–30
31	40	–48
78	95	17
215	214	236

167	283	309	434	512	608	274	713	82	903
280	406	472	159	296	152	460	264	39	–75
349	137	563	618	–947	591	–958	380	65	–829
265	459	128	–703	613	235	–215	521	–721	–6493
420	261	–796	–841	308	–764	309	–849	–9604	146
593	–825	204	397	247	318	132	–706	76	2704
178	–570	–485	204	105	–973	341	–693	148	92
–837	–968	–607	–972	430	–896	–604	–948	–85	–98650
–672	309	815	120	–867	409	–897	–874	–689	–3518
–905	147	937	584	–698	240	–786	150	–4053	637
–159	–361	1546	0	–1	–80	1944	2036	–14682	105063

45	63	87	31	56	94	26	79	70	12
78	95	60	19	27	70	98	16	45	60
30	59	31	40	68	19	–20	84	61	85
59	–80	48	95	90	51	–57	40	13	–32
61	–43	24	–87	31	35	41	62	90	–78

96	78	12	–28	47	60	93	97	24	45
20	51	39	30	85	89	–65	23	57	90
83	–27	50	72	43	46	10	95	38	76
52	82	67	16	10	18	74	60	47	–53
17	90	29	–27	24	42	58	35	96	10
93	–64	15	20	93	37	–89	21	80	34
40	–37	98	65	51	68	–13	74	12	–89
62	10	54	–43	80	27	30	83	35	–71
81	46	70	–64	69	50	76	18	26	94
74	12	36	98	75	23	42	50	89	67
891	335	720	207	846	729	304	837	783	250

359	9535	604	413	2190	4157	516	8594	740	1507
7569	174	2895	390	647	823	7082	250	6294	960
408	812	731	706	574	496	395	913	103	8023
163	3720	1048	6054	1697	273	279	–341	857	649
914	–647	269	318	481	5082	153	–7082	5019	752
792	–1093	817	8249	156	761	6042	361	431	836
5021	356	4508	–634	3078	–845	821	2473	520	6395
8630	2680	932	–267	423	–3978	3150	749	648	548
325	906	126	–9835	310	–109	264	605	3976	125
206	185	958	142	962	634	4987	128	837	409
127	–263	9053	5061	835	2014	730	9086	2719	257
4381	–8472	316	975	5289	361	674	–597	301	9168
948	–598	470	–7529	206	–580	968	–1632	4658	314
6057	704	3749	–807	4058	–6792	401	–805	285	7431
874	419	527	182	739	905	5893	467	961	270
36774	8418	26775	3418	21645	3202	32355	13169	28349	37674

Multiplication

243	906	795	545	490	808	246	338	215
×527	×930	×889	×701	×770	×847	×481	×632	×856
206064	142242	735375	243070	446880	470256	186468	56108	59340

782	775	980	764	909	515	863	465	302
×527	×930	×889	×701	×770	×847	×481	×632	×856
412114	720750	853440	535564	699930	436205	415103	293880	258512

808	862	597	212	695	817	971	998	665
×315	×598	×296	×352	×336	×585	×622	×459	×383
254520	515476	176712	74624	233520	477945	603962	458082	254695

704	489	651	787	936	617	284	460	199
×972	×303	×694	×419	×130	×128	×392	×571	×287
684288	148167	451794	329753	121680	78976	111328	262660	57113
238	146	421	311	174	732	983	863	871
×142	×607	×888	×801	×397	×370	×743	×857	×978
33796	88622	379176	249111	69078	270840	730369	739591	851838
848	912	293	197	630	785	922	122	666
×413	×595	×676	×249	×169	×346	×466	×526	×992
350224	542640	198068	49053	106470	271610	429652	64172	660672
691	795	621	735	531	132	575	191	210
×201	×453	×952	×375	×261	×154	×987	×463	×993
138891	360135	591192	275625	138591	20328	567525	88433	208530
807	931	250	959	931	326	374	364	579
×596	×237	×510	×133	×506	×153	×292	×441	×934
480972	220647	127500	127547	471086	49878	109208	160524	540786
320	618	799	191	222	525	422	910	385
×247	×610	×360	×424	×379	×464	×30	×205	×754
79040	376980	287640	80984	84138	243600	265860	186550	290290
942	333	477	356	469	267	613	402	751
×490	×256	417	454	×668	×819	×668	×863	×631
461580	85248	198909	161624	313292	218673	409484	346926	473881
550	773	999	605	871	306	905	632	923
×616	×959	×508	×364	×626	×567	×851	×322	×631
338800	741307	507492	220220	545246	173502	770155	203504	582413
142	436	271	562	390	824	248	110	479
×646	601	×396	×467	×669	×872	×435	×827	×528
91732	262036	107316	262454	260910	718528	107880	90970	252912
485	631	133	676	180	134	452	677	302
×449	×436	×251	×762	×453	×104	×788	×623	×341
217765	275116	33383	515112	81540	13936	356176	421771	102982

525	361	512	101	123	777	565	839	131
×147	×798	×329	×423	×111	×775	×924	×409	×790
77175	288078	322048	42723	80253	602175	522060	343152	103490

776	803	946	174	222	334	424	179	357
×159	×435	×657	×865	×518	×570	×495	×470	×736
123384	349305	621522	150510	114996	190380	209880	84130	262752

448	667	371	400	885	694	199	936	320
×179	×890	×587	×431	×825	×358	×164	×401	×525
80192	593630	217777	172400	730125	248452	32636	375336	168000

178	231	555	907	419	558	172	179	224
×715	×151	×500	×957	×343	×391	×314	×316	×689
127270	34881	277500	867999	143717	218179	54008	56564	154336

726	612	955	429	106	926	269	967	119
×678	×773	×750	×636	×126	×880	×704	×864	×528
492228	473076	716250	272844	13356	814880	189376	835488	62832

916	661	703	285	278	306	185	399	604
×945	×726	×133	×574	×867	×179	×974	×796	×783
865620	479886	93499	163590	241026	54774	180190	317604	472932

333	880	225	525	612	278	531	333	131
×970	×180	×469	×479	×493	×768	×726	×332	×834
323010	158400	105525	251475	301716	213504	385506	110556	109294

Multiplication

4×2=8
509×2=1018
8105×9=72945
25×15=375
876×8=7008
2619×3=7857
45×8=360
623×4=2492
5980×6=35880
73×4=292

258×9=2322
3467×5=17335
52×2=104
480×6=2880
9024×9=81216
38×7=266
942×5-4710
2751×2=5502
94×3=282
715×7=5005

2647×6=21882
15×6=90
301×4=1204
8502×4=38008
86×9=774
697×3=2091
6138×3=18414
67×5=335
7832×6=46992
8420×9=75780

21×8=168
4096×7=28672
4975×2=9950
79×6=474
6348×8=50784
1769×8=14152
134×3=402
1573×4=6292
7081×4=28624
2153×2=4306
5304×5=26520
8296×7=58072
8296×7=58072
24×88=2112
37×98=3626
851×99=84249
187×53=9911
245×21=5145
309×19=5871
408×38=15504
561×60=33660
620×42=26040
716×90=64440
832×57=47424
954×74=70596
973×86=83678
1725×51=87975
2698×24=64752
3980×30=119400
4509×65=293085
5062×73=369526
6874×68=467432
7431×80=594480
8146×72=97752
9357×49=458493
8230×97=798310
92854×84=7799
8240×568=4680
73041×90=6573
60378×16=966048
51762×27=1397
47609×70=3332630
30427×32=973664

29185×63=1838
18596×45=836820
54930×81=4449
848×276=234048
854×965=824110
902×804=725208
627×108×67716
105×519=54495
570×843=480510
489×751=367239
616×397=243361
236×632=149152
791×420=332220
942×455=466290
839×457=383423
723×980=708540
680×134=91120
508×268=136144
417×873=364041
396×629=249084
204×316=64464
165×501=82665
751×702=527202
1375×562=772750
2610×148=386280
3784×625=2365000
4208×201=845808
5429×874=4744946
6057×903=5469471
7906×417=3296802
8591×730=6271430
9832×986=9694352
4163×359=1494517
4097×238=975086
5638×149×840062
14902×52=774904
7105×0.098=696.29
9674×603=5833422
63025×7.64=483.23
853×4.017=3426.501
0.3081×0.926=0.285
2984×351=1047384

0.2176×87.5=19.04
9108×379=3451932
8240×568=4680320
7894×740=5841560
6372×953=6072516
5423×182=986986
4617×194=895698
3581×807=2889867
2056×628=1285000
1905×401=763905
3769×236=889484
2647×3740=9899780
3068×2698=8277464
9854×7219=71136026
1370×4805=65828508
8401×6457=54245257
4936×9523=47005528
6125×5184=31752000
2519×8306=20922814
7093×1962=13916466
5782×3071=17756500
169×987=166803
408×996=406368
3600×0.92=3312
405×123=49815
34×1.2=40.8
32×0.4=12.8
98×0.32=31.36
32×0.04=1.28
98×0.032=3.136
32×0.004=0.128
98×0.0032=0.3136
2594×376=975344
4608×0.189=871
7832×897=7025304
94120×6.4=602368
029×738=5925402
975×45.12=43992
5176×0.625=3235
3061×903=2764083
6843×201=1375443
6549×643=4211007

Dividing

72/2=36
135/5=27
246/3=82
126/9=14
413/7=59
300/4=75
744/8=93
288/6=48
183/3=67
435/5=87
8829/9=981
420/4=105
2634/6=439
752/2=376
4336/8=542
4050/5=710
1906/2=953
2124/3=708
1068/4=267
4368/7=624
402/2=201
6776/7=968
3282/6=547
2430/3=810
5536/8=692
1380/4=345
3180/6=530
7101/9=789
3408/8=426
865/5=173
3896/4=974
966/7=138
1184/2=592
4830/6=805
807/3=269
2250/5=450
2468/4=617
682/2=341
5648/8=706
7407/9=823
8/2=4
375/15=25
49815/123=405
40.8/1.2=34
12.8/0.4=32
31.36/0.32=98

1.28/0.04=32
3.136/0.032=98
960/24=40
810/45=18
7505/79=95
6640/80=83
5920/16=370
4080/68=60
3127/53=59
2160/30=72
1152/72=16
2184/91=24
63207/90=702.3
533484/87=6132
420616/74=5684
113148/63=1796
278772/52=5361
85075/41=2075
366873/39=9407
98504/28=3518
72885/15=4859
214240/26=8240
9724/26=374
8151/13=627
7739/71=109
62560/80=782
5556/12=463
49572/54=918
30150/67=450
23128/98=236
17535/35=501
43855/49=895
379426/42=9034
706860/85=8316
235662/31=7602
658145/97=6785
406164/68=5973
87362/19=4598
115710/30=3857
55637/23=2419
94240/76=1240
325134/54=6021
636768/792=804
35984/208=173
194394/537=362
403425/815=495

86163/373=231
531340/620=857
934832/984=948
286090/469=610
363726/501=76
74314/146=509
94235/401=235
87040/256=340
752128/832=904
64220/380=169
548784/927=592
41748/603=716
385746/478=807
219537/519=423
107912/164=658
620895/795=781
553149/6829=81
273375/3645=75
97351/1453=67
761664/7934=96
81018/4501=18
84942/2178=39
424901/8017=53
120048/5002=24
611640/6796=90
394044/9382=42
937015/965=971
0.08988/6.42=0.014
0.070654/0.136=0.520
63366/708=89.5
55.426/214=0.259
415473/591=703
315.333/45.9=6.87
280932/82=3426
17.316/0.037=468
241893/7803=31
99/0.368=269
83619/27=3097
71967/149=483
649612/7061=92
560/0.875=640
415693/593=701
33154/60.5=548
2485/2.84=875
122563/901=136
54/0.432=125

124352	479272	129780	111252	278502	107568	36232	4280	170937	108058
÷464	÷862	÷140	÷292	÷399	÷216	÷56	÷20	÷487	÷194
268	556	927	381	698	498	647	214	351	557
11356	261611	38745	31758	65637	183816	148740	778872	5082	361045
÷34	÷931	÷105	÷67	÷297	÷888	÷670	÷996	÷154	÷443
334	281	369	474	221	207	222	782	33	815
3172	409041	183918	554302	680200	415454	146520	153945	864152	525980
÷61	÷967	÷453	÷578	÷760	÷551	÷220	÷933	÷991	÷595
52	426	406	959	895	754	666	165	872	884
276744	319440	81545	262224	864708	707256	25632	358820	102678	75047
÷887	÷528	÷238	÷432	÷897	÷893	÷36	÷466	÷471	÷497
312	605	347	607	964	792	712	770	218	151
19398	702400	221341	82904	274872	261208	220866	389456	173052	818573
÷122	÷878	÷569	÷86	÷312	÷412	÷393	÷964	÷759	÷983
159	800	389	964	881	634	582	404	228	731
83226	50836	95832	339586	210148	83895	315704	385436	66788	284499
÷194	÷179	÷121	÷481	÷428	÷105	÷589	÷577	÷118	÷369
429	284	792	706	491	799	536	668	566	771
176148	236880	946153	117603	186744	672084	24710	34980	179816	199420
÷699	÷336	÷997	÷219	÷372	÷882	÷706	÷66	÷364	260
252	705	949	537	502	762	35	530	494	767
57600	63544	313780	21973	78625	4896	179056	29085	16500	78150
÷200	÷94	÷580	÷301	÷85	÷48	÷589	÷277	÷50	÷521
288	676	514	73	925	102	304	105	330	150
654477	574443	159045	223136	176916	6401	326893	675234	102261	67614
÷811	÷747	÷345	÷367	÷276	÷37	÷697	÷966	÷383	÷573
807	769	461	608	641	173	469	699	267	118
29445	355047	220671	207545	74025	80032	249228	613137	304764	313028
÷151	÷609	÷297	÷403	÷329	÷244	÷514	÷903	÷466	÷556
195	583	743	515	225	328	483	679	654	563

42897	533820	27224	324952	213444	256846	1110	18675	107751	56518
÷79	÷651	÷82	÷604	÷539	÷334	÷30	÷225	÷733	÷154
543	820	332	538	396	769	37	83	147	367

470400	140896	111418	337648	201600	440360	36616	512064	534462	257640
÷800	÷259	÷986	÷752	÷900	÷872	÷184	÷672	÷562	÷458
588	544	113	449	224	505	199	762	951	565

279734	266976	195146	453500	242003	28002	28059	661168	78080	108030
÷754	÷324	÷526	÷907	÷271	÷718	÷141	÷688	÷640	÷277
371	824	371	500	893	39	199	961	122	390

8550	189202	233996	133342	141882	43792	28737	22388	454530	862875
÷57	÷383	÷274	÷209	÷221	÷322	÷93	÷29	÷834	÷975
150	494	854	638	642	136	309	772	542	885

30668	469017	108324	86457	5028	408996	444744	536928	16170	425904
÷82	÷783	÷918	÷483	÷419	÷756	÷696	÷987	÷294	÷467
374	599	118	179	12	541	639	544	55	912

88888	107616	260887	62516	248584	62906	135040	7475	438060	790830
÷542	÷354	÷641	÷91	÷386	÷886	÷640	÷575	÷490	÷870
164	304	407	676	644	71	211	13	894	909

292989	204512	42642	420210	257744	63688	591595	239547	445556	11946
÷381	÷913	÷309	÷805	÷358	÷419	÷661	÷357	÷668	÷22
769	224	138	522	718	152	895	671	667	543

480430	199670	100672	13114	615479	41724	678878	153996	62048	512686
÷898	÷410	÷416	÷166	÷701	÷549	÷697	÷939	÷224	÷887
535	487	242	79	878	76	974	164	277	578

803440	198541	177480	25935	402000	35375	94024	129048	174124	92625
÷830	÷791	÷870	÷95	÷750	÷283	÷584	÷152	÷404	÷195
968	251	204	273	536	125	161	849	431	475

183913	95040	372600	129696	364266	238933	443515	522759	4602	214830
÷521	÷440	÷900	÷386	÷413	÷337	÷535	÷643	÷354	÷682
353	216	414	336	882	709	829	813	13	315

8400	511215	174087	311140	758620	43099	618192	508796	86292	707296
÷14	÷985	÷841	÷470	÷830	÷917	÷848	÷622	÷423	÷992
600	519	207	662	914	47	729	818	204	713
496620	362700	503041	450240	358360	491682	6916	59055	42840	281649
÷801	÷930	÷973	÷480	÷620	÷722	÷91	÷127	÷68	÷421
620	390	517	938	578	681	76	465	630	669
516405	14430	63457	93840	10716	40680	16072	19038	20922	19630
÷597	÷30	÷713	÷408	÷12	÷452	÷82	÷167	÷66	÷755
865	481	89	230	893	90	196	114	317	26
15498	155067	27063	388620	235410	495075	407862	330052	429942	659634
÷189	÷407	÷93	÷508	÷885	÷615	÷581	÷436	÷547	÷986
82	381	291	765	266	805	702	757	786	669
900736	785850	96669	240012	452488	61952	144552	139860	104640	142444
÷908	÷930	÷207	÷531	÷652	÷352	÷456	÷740	÷872	÷478
992	845	467	452	694	176	317	189	120	298
66234	83400	784696	240340	20821	287500	85888	100498	151498	127576
÷114	÷149	÷964	÷305	÷443	÷625	÷976	÷218	÷318	÷431
581	560	814	788	47	460	88	461	477	296
14136	19654	203175	179778	425600	102600	341502	320166	395604	634400
÷213	÷634	÷675	÷498	÷800	÷228	÷987	÷693	÷396	÷650
672	31	301	361	532	450	346	462	999	976
681472	84400	122360	102900	201348	18990	338519	159225	504400	258810
÷968	÷422	÷437	÷150	÷564	÷30	÷673	÷975	÷520	÷421
704	200	280	686	357	633	503	473	970	610
7848	151980	6831	157464	432003	483678	9384	547230	26052	268695
÷436	÷298	÷69	÷324	÷741	÷689	÷69	÷555	÷52	÷315
18	510	99	486	583	702	136	986	501	853
340952	741198	143136	87174	206910	192685	86180	208656	380240	142003
÷872	÷943	÷284	÷261	÷495	÷445	÷278	÷504	970	÷211
391	786	504	334	418	433	310	414	392	673